DEVELOPING A SCHEME OF WORK

for **Primary Mathematics**

Sue Atkinson

Series Editor: Shirley Clarke

Hodder & Stoughton

A MEMBER OF THE HODDER HEADLINE GROUP

This book is dedicated to the memory of Hilary Shuard who spent her working life seeking to improve mathematics teaching in schools.

British Library Cataloguing in Publication Data

A catalogue record for this title is available from The British Library

ISBN 0 340 62062 5

First published 1996
Impression number   10 9 8 7 6 5 4 3 2 1
Year   1999   1998   1997   1996

Typeset by Multiplex Techniques, Orpington, Kent.
Printed in Great Britain for Hodder & Stoughton Educational, a division of Hodder Headline Plc, 338 Euston Road, London NW1 3BH by The Bath Press, Avon.

DEVELOPING

A SCHEME OF WORK

for **Primary Mathematics**

CONTENTS

ACKNOWLEDGEMENTS

A great many schools and teachers contributed towards the writing of this book – many of them without knowing it as I worked with them as a teacher or as an INSET provider. I particularly want to thank Mary Walker and many other Open University students I worked with. It was my pleasure to participate in their maths curriculum development over the years and to share with them in finding effective ways to improve the teaching of maths in their schools. Many people gave their advice and support in a variety of ways: Peter Clarke; Muriel Chester; Hilary Evens; Champa Bryson and the staff of Bessemer Grange Junior School, Southwark; Jean Andrews and Dylan Lodge and the other staff at Keyworth Primary School in Southwark; the staff of Blessed Sacrament School, Islington; the staff of Windmill First School, Oxford; the staff of West Oxford Primary School; Sheila Ebbutt; Fran Mosely; Grace Cook and many others on BEAM courses about developing a scheme of work.

This book has taken several years to complete, and I want to thank Elizabeth Wright at Hodder and Stoughton for her patience and encouragement and Shirley Clarke for her helpful and creative editing, and her practical advice. Their cheerful support enabled me to keep going even when things became tough. We hope we have made this book accessible and clear for any school or individual teacher as they work on putting together and using their scheme of work.

Sue Atkinson

INTRODUCTION

How to use this book

This section includes:
- an overview of what the book is about;
- some starting points for using the book.

The book is split into five sections as shown on the contents list:

A An overview of the scheme of work and the people involved.
B Creating the scheme of work.
C Planning for the scheme of work.
D Getting started on the scheme of work.

To get started you could:

- look at the contents list and find the sections you need;
- dip into the bits that seem relevant to your situation (there are cross references to other relevant parts of the book to help your 'dipping in');
- pick out the bits which relate to your school development plan or which you want to work at as a staff or individual;
- look at section D, 'Getting started on the scheme of work'.

Each chapter starts with a box that shows the contents of that chapter. This will help you in selecting chapters to meet your particular needs.

The middle part of the chapter shows a variety of ways in which you might work towards achieving what is described in that section. This part contains:

- 'how to do it' bits and explanations of why that is a way to do it. These 'why' sections are the theories of mathematics education – in other words, what we know at the moment about how children learn maths;
- examples from many schools and maths co-ordinators who have been involved in gathering the data for the book either directly or indirectly over a few years, so the examples are from a very wide range of schools and children;
- what might happen, including things that might go wrong.

It is absolutely crucial to understand that things *will* go wrong! Problems and conflicts arise in any situation in which people try to work together to improve what they do. Conflict and difficulty are not signs of failure. They are indications that people are thinking about what they are doing and trying to find a way to improve their teaching.

The final part of each chapter has an action box with some suggestions of what you might do as a response to the chapter. These can be used as a basis for INSET in your school or cluster.

INSET   There is also an INSET icon beside some of the bits of the main text where these could form the basis of INSET meetings or work between meetings.

- Some of the INSET activities at the ends of chapters will be quite easy, and they are things that you could do on your own and do fairly quickly – today, or tomorrow when you are with your class. You could select two of your best maths activities to share at the next staff meeting; get all of your maths equipment out of the cupboard to show others what you have; tape some children doing their maths; do a mental maths session with your class and write down what you asked them and share this with a colleague.
- Other things are harder; you might need a week or a month and you might want to consult other colleagues. This could be deciding on a starting point for getting your scheme of work together and deciding on priorities.
- The final group of actions are things that are your long-term objectives. These might take a year or more to get going and might well need collaboration with others. Getting your scheme of work into some recognisable form as a scheme of work might be in this category.
- Some of these activities can be done with photocopied sheets from the appendices or others in the text.

A NOTE ON TERMINOLOGY

An important part of the maths curriculum is the part that is about the processes of mathematical thinking. This was often called 'AT1' in the English and Welsh curriculum and it includes the problem solving and enquiry of the Scottish 5–14 curriculum. In this book, I will call this 'using and applying maths'.

> - Look through the book and decide where you could start. (See contents list at the front of the book and look at the start of each chapter for its key contents.)
> - Talk with a colleague about possible ways that the two of you might work together in the early stages.
> - Consider starting by asking the head if you could work with him or her on some kind of activity to establish what is happening in maths in the school at the moment. (See the questionnaire on page 130.)

WARNING

Most schools that work on their scheme of work report that they took on too much, so:

- think small;
- be realistic about what you can do in the time you have available;
- do a few parts of the scheme of work well rather than rush it all (e.g. you could have a whole school focus on number for half a term and follow this by half a term on shape and space, maybe keeping the number ticking over).

Many of us have been through huge changes in the past few years, so find some tactics to share the work load.

REMEMBER

What you end up with written down is very important (you must demonstrate that you are covering the curriculum), but what goes on in the classroom is much more important.

Developing a scheme of work is difficult, time consuming and at times utterly frustrating. So we must be realistic and keep our main aim in our mind, i.e. to work towards improving the learning situation for the child. Nothing must get in the way of that.

SECTION

AN OVERVIEW OF A SCHEME OF WORK AND THE PEOPLE INVOLVED

This section is divided into two chapters:

CHAPTER 1
An overview of the mathematics scheme of work

CHAPTER 2
Roles within the school

CHAPTER **1**

AN OVERVIEW OF THE MATHEMATICS SCHEME OF WORK

This chapter includes:
1 What is a mathematics policy?
2 What is a mathematics scheme of work?
3 What is the purpose of the scheme of work?
4 How the scheme of work relates to the National Curriculum
5 Interpreting the Programmes of Study

WHAT IS A MATHEMATICS POLICY?

Our finished scheme of work will almost certainly have what we might call a policy (the rationale or philosophy that underlies what we want to do and why) attached to it. These 'whys' are important, and a few pages of this before the nitty-gritty of the scheme of work is quite important. In this book, I have woven these aims into each chapter so that the 'this is what to do' bits have explanations with them saying why they are important.

Your school maths policy starts something like this:

Our school policy on why we teach maths the way we do.
Date: ......
Compiled by: ...... , ... , ... and ......

One example of a school policy document is shown in figure A1. Yours will need to fit your school and might be very different.

You might want to link your maths policy with the wider aims of the school and have a maths mission statement, for example:

BARROW HOUSE SCHOOL MISSION STATEMENT ON TEACHING MATHS
In Barrow House School, we aim to give every child an equal entitlement to a broad and balanced maths curriculum. We believe that every child, etc.

Maths Policy

1 Teachers will plan on a long-term, one- to two-year basis. They will plan collaboratively within the team. Maths activities will occur on a daily basis within each class in line with this planning.

2 There will be regular review of these plans to ensure balance, breadth, continuity and progression of the mathematics curriculum, which will be based on the National Curriculum Programmes of Study.

3 Mathematics offers a way of analysing and synthesising our experiences through the acts of describing, organising, explaining and predicting in order to make sense of the real world.

4 We feel that maths provides children with experience to think logically and deal with abstract concepts and skills that can be used across the whole curriculum and in other learning experiences, such as art, science, DT, IT, history and geography.

5 Learning experiences provided by the teacher should have a balance between investigative work, practical problem solving and 'pure' mathematical activities.

6 We aim to nurture positive attitudes by matching the task to the child. We feel that successful learning enables children to develop the confidence to meet the challenge of new work.

7 To ensure a coherent approach to each child's learning, account is taken of what they already know.

8 Children's progress is monitored on a regular basis through informal teacher assessment and the ongoing mathematics record provided for each child. Samples of work with explanatory notes are kept to show evidence of achievement and to facilitate planning. The PLR is used and a report is made available to parents.

9 We aim to offer a non-sexist and non-racist maths curriculum which is appropriate and accessible to the needs and abilities of all our pupils.

10 We aim to group children in a wide range of ways, from an individual approach to a large group situation, depending on the particular needs of the child and/or the task.

11 We aim to present an agreed approach to the notion of good practice and the nature and type of teacher intervention.

12 Senior management and the maths co-ordinator will endeavour to support and develop the maths teaching skills of staff through various INSET programmes within school and at other venues.

13 There will be a wide range of appropriate mathematical equipment readily available for all children. These materials should be clearly labelled and organised to ensure easy storage, access, upkeep and replacement.

14 There is a central fund of high-quality published mathematics resources for all teachers to use and share with the children and parents.

15 Displays are a way of valuing children's mathematical achievements. They are an important learning resource and should reflect current work in progress, positive images and cultural diversity.

16 We aim to encourage parents as active partners in the development of their child's maths skills. We recognise and aim to utilise the valuable source of learning parents and others in the local community have to offer.

17 The needs of the school as outlined in the school development plan should be kept under review, and common and agreed approaches towards planning activities, the learning environment and time management should be pursued.

18 This policy will be reviewed annually. Next review due on ....

Signed by ....   Date ....

FIGURE A1 *Example of a school maths policy document. Many thanks to Jean Andrews and the staff of Keyworth Primary School, Southwark, for this maths policy.*

You might have some paragraphs that outline your school philosophy on such topics as equal opportunities, etc. and relate these to teaching maths. Then this might be followed by a list of aims such as those below.

OUR AIMS AND PURPOSES
- To help every child to be a confident mathematician.
- To enable every child to enjoy their maths.
- To help every child to have the confidence to tackle any problem that they might come across in their daily life.
- To give children a firm basis of knowledge and skills so that they are numerate and able to work flexibly and think clearly.
- To enable children to work as a part of a group to find appropriate strategies for problem solving.
- To give children the chance to achieve at the highest level that they are capable of.
- To give every child access to stimulating and appropriate maths activities, chosen from a broad and balanced curriculum, that help children to think and to enjoy what they are doing.

As you work together to write your scheme of work and policy statement, you will find that you can add to and refine your list of aims. The list in figure A2 might help your discussions. It is the list that I have kept in mind whilst working with teachers during the writing of this book.

Strategies for action

Following on from your list of aims, you might have a section called 'strategies for action'. These can be the strategies that fit the ways in which you teach maths in your school to fulfil the aims of the policy. These might include such things as:

- Teachers will plan together at the start of each half term and each two weeks throughout the term.
- Work will be planned on the planning sheet; one copy should be placed on the classroom wall to inform parents and another in the file in the office at the start of each two-week period.
- Any ability groups for maths will be flexible and reviewed at least each half term.
- Resources from the central store will be booked out and signed for. These are to be returned at the end of that maths topic.
- Each class will have ongoing maths as a part of the plan (e.g. number, information handling, Logo and other computer work).

The kinds of issues that your policy will need to cover are:

- the nature of maths (how does maths contribute to the child's school experience, the aims and objectives of the maths curriculum (discussed throughout this book) and the need to relate your aims in maths to the overall aims of the school);
- how the school policy matches the National Curriculum (page 28);
- pupils' mathematical experiences and activities (i.e. providing a balance of learning experiences (pages 35 and 42);
- children's recording of their work (pages 32);
- maths and cross-curricular issues (applying maths and pure maths, pages 77);
- assessment and record keeping (pages 111 – 23);
- staffing and resources (pages 88 – 108);
- classroom management (page 140);
- continuity and progression (page 60);
- equal opportunities (page 161);
- new technology (page 96);
- ways that you will evaluate what you do (page 162);
- ways that you will encourage individuality amongst teachers as they actually teach the maths.

See the Non-Statutory Guidance for a further discussion of these points.

FIGURE A2 *Things to consider when putting together a maths scheme of work.*

- Each class will have a maths topic that all children work on for a few weeks, e.g. two-dimensional shape, symmetry, using data handling packages on the computer, circles, place value, multiplication, angles.
- The floor robot will be given to each class for two weeks at a time. The timetable for this will be posted at least a term ahead on the notice board and staff should ensure that the robot is in use most of the time. Space needs to be given to this within the classroom, perhaps by closing down some other activity for those two weeks (e.g. remove one painting easel, or the shop, or put big building bricks into the corridor).
- The construction equipment will be sent to each class on a rota for two weeks. Please book your weeks for these kits on the notice board.
- Each teacher will use the published maths scheme for consolidation and as a structure to base work on, but they are encouraged to use the other resources in the staff library to supplement the scheme. No child should be expected to work only from the scheme.
- Maths evaluation meetings will be held regularly. (If you actually fix the date of the next one in your diary, it is less likely to get overlooked in the day-to-day business.)

Strategies for action will obviously vary from school to school, and each year group (or other planning group) may well want their own list that they develop from the main school list.

These strategies are very important. They are the link between the philosophy and the practice. If you don't have them, all that high-flown stuff in the policy could fail to influence practice in any way!

WHAT IS A MATHEMATICS SCHEME OF WORK?

The single most important thing about a scheme of work is that it is something that will go on changing and developing. It is not a finished and polished document that sits on the shelf and gathers dust. The scheme of work outlines what maths will be taught and how and when.

- It is the main focus of what we are planning in our maths.
- It is the way in which inspectors, parents and governors can assess if we are doing our job adequately.
- It is our guide to ensuring that we are meeting our legal requirements and teaching the National Curriculum.
- It is our guide to ensuring that we are giving every child in our class the same opportunity to be educated and to achieve their maximum potential.

That is all quite serious stuff! So when we are engaged in writing or amending our scheme of work, we are doing something of great importance.

Your scheme of work will probably be different from all the examples of schemes of work in this book and the schemes of work produced by schools near you. Your school is unique, with your catchment area, your particular set of resources and your staff. It needs to work for you, so although there are many examples in this book, pick and choose and adapt for what you want.

There are a number of ways in which you could present your scheme of work, and several examples are given in this book, but one way could be to have columns as in figure A3 with space for notes. Everyone would have a copy, and it would be the basis for maths in the classroom.

You might want to have some of your scheme of work organised by age groups, as this can make it simple to operate. (But note that it is important that a child's maths tasks are decided by the child's ability, not by some predetermined idea for what, say, an eight-year-old should be doing.) Some schools do this just as Key

Stages 1 and 2, and others do a separate scheme of work for each year group (so year 3 (see Figure A4) would have some very early work to be covered by the less able that would be separate from the scheme of work for the year 1 and 2 children, but would also include decimals, fractions and rotational and reflective symmetry that are a part of the year 4 and 5 scheme of work).

X School scheme of work Version X Date: .... Compiled by:			
Programmes of Study	Activities	Resources	Notes
• number • reading and ordering numbers to 1000	• mental maths – counting in 10s, 100s 1000s • patterns on 100 square • 'hops' on the number line • using large numbers scheme page 14–18 • what's my number? page 19–21	• Unifix 100 line • wall number line (extended) • 100 squares	
• use calculator to explore numbers	• BEAM calculator book activities – – – • hundreds and thousands activity scheme page 30–33	• calculators • BEAM book	

FIGURE A3 *Scheme of work, format 1.*

YEAR 3

NUMBER

i) UNDERSTANDING

Consolidate understanding of numbers to 100 – use of Base 10 blocks, 0–100 cards, number track/line, number square.

Exchange games involving 'tens' and 'units'. Grouping and exchanging t.u.

Applications to money up to 100p and length up to 100cm.

Read, write and order numbers to at least 1000.

Grouping and exchanging h.t.u. through games and other activities.

Recognising the importance of first digit and the value of all digits.

Understand 'zero' as a place holder.

Round up or down to the nearest 10.

Use decimal notation to record money; show this in written form, on a calculator and on a decimal abacus.

Consolidate knowledge of 'halves' and 'quarters'.

Understand meaning of a 'third' and other simple fractions.

ii) OPERATIONS

Know and use addition and subtraction facts to 20 including zero.

Use a variety of approaches and apply these facts to everyday situations.

Add 10, 20, 30, etc. Show addition patterns on a 100 line/square.

Add 'tens' and 'units' with and without exchange up to 100.

Subtract 'tens' and 'units' with and without decomposition.

Use practical aids – Base 10 blocks, number lines, money, calculators.

Understand 'taking away' and 'finding the difference'.

Encourage mental methods for +n and –n. Talk about these.

Use addition squares up to 10 by 10 and 20 by 20.

Understand the concept of 'equal sets' and repeated addition/subtraction leading to multiplication and division.

Understand sharing situations leading to 'equal sets' with and without remainders.

OPERATIONS Cont'd

Understand about multiplication arrangements and arrays.

Revise knowledge of multiplication facts up to 5 x 5; use multiplication square 5 x 5.

Extend knowledge of multiplication to all facts in the 2, 5 and 10 times table.

Apply knowledge of 4 rules of number to everyday problems, especially money.

ALGEBRA

i) PATTERNS & RELATIONSHIPS

Number of patterns and relationships linked with number bonds.

Creating number sequences – simple rules.

Further work on odd and even numbers.

Counting patterns in 2s, 5s and 10s.

Explore square and rectangular numbers practically – link with multiplication.

ii) EQUATIONS & FUNCTIONS

Missing number sentences/empty box arithmetic – variety of activities.

Understand the use of a symbol to represent a number.

Mappings and arrow rules.

Simple INPUT and OUTPUT machines.

iii) CO-ORDINATES & GRAPHICAL REPRESENTATION

Using co-ordinates on a square grid. Understand horizontal and vertical and how to label the two axes.

Understand the convention for naming a 'square' on the grid, using an 'ordered pair'. Activities with pictures, patterns and shapes.

YEAR 3

MEASURES

LENGTH Measure in metres, 1/2m and 1/4m. Introduce use of centimetres and decimetres. Use notation cm, dm, m.
Know that 100cm = 1m; 50cm = 1/2m; 25cm = 1/4m.
Use a 30cm ruler and understand how to use a ruler correctly.
Use a tape measure, metre stick, trundle wheel.
Estimate to nearest m or cm.

WEIGHT Conservation of weight activities.
Activities using kilos and 1/2 kilos; Weigh using 100g and 50g weights.
Show weight using a stretched rubber band.
Practical activities using (a) a two-pan balance (b) a dial kitchen scale.
Estimate in kilos and 1/2 kilos; know of pounds and ounces; stones and pounds.

CAPACITY AND VOLUME Conservation of volume activities with liquids and solids.
Measure capacity in litres, 1/2l and 1/4l. Use a wide range of jugs, cylinders, bottles and other everyday empties. Estimate and compare using litres and 1/2l.
Use cubes – multilink, wood, plastic – to find capacity of cube and cuboid boxes.
Introduce the cm cube and fill small boxes to find their capacity. Estimate using cubes. Know of pints and gallons.

AREA Find area by covering surfaces with regular shapes and counting.
Understand tessellations and tiling patterns in relation to area – everyday examples. Find area by counting squares – use large squares then introduce small ones, and compare the results.
Conservation activities – cut shape, rearrange pieces.

TIME Know relationships 60min = 1hr; 30min = 1/2hr; 15min = 1/4hr; 45min = 3/4hr. Revise telling the time using o'clock, 1/4 past, 1/2 past and 1/4 to – link these with 0, 15, 30 and 45 minutes 'past' in digital time. Consolidate understanding of the clock face – intervals in hours and in minutes. Counting patterns. Begin using 5 min intervals in telling time. Use dials showing days, numbers and months showing the cyclic nature of time. Use to show date. Construct calendar grid and time line. Create and use own timing devices e.g. sandclock, candle clock.

SHAPE AND SPACE

ANGLE Understand right and left turns, whole 1/2 and 1/4 turns – using themselves and everyday objects.
Make a right angle by (a) folding paper (b) cutting a card L shape.
Know the meaning of horizontal and vertical.
Identify right angles on 2-D, 3-D shapes and in the environment.
Routes and mazes – give and follow instructions with right-angle turns.
Use the four major points of the compass.

SHAPE & SPACE Construct 2-D shapes by drawing and cutting, folding and cutting, with straws and pipe cleaners/geostrips.
Know the names of shapes. Use 2-D shapes in patterns, mosaics, tessellations. Explore which shapes tessellate.
Make 3-D shapes using plasticine, polydron, straws and pipecleaners.
Explore properties of 2-D and 3-D shapes by counting corners, edges, faces.
Recognise, name and collect 2-D and 3D shapes from the environment. Make repeating patterns along a line (translation).
Investigate reflective symmetry by folding activities and mirrors work.
Make rotation patterns i) by pinning a shape and turning slowly (stop and draw) ii) using two hinged mirrors.
Recognise reflective and rotational symmetry in objects and pictures.

HANDLING DATA

Sort and classify objects using one or two criteria.
Record on Venn, Carroll and Tree sorting diagrams.
Collect data from surveys, class lists, games scores.
Record using tallying methods and frequency tables.
Construct block graphs to display data – talk about and interpret these.

FIGURE A4 Scheme of work, format 2.
Thanks to Muriel Chester, Champa Brysen and the staff of Bessemer Grange Junior School, Southwark, for this part of their scheme of work.

WHAT IS THE PURPOSE OF THE SCHEME OF WORK?

A scheme of work is:

- for using;
- for talking about;
- for jotting notes on;
- for amending;
- a guide for improving the child's learning experience;
- ensuring that we are meeting statutory requirements;
- helping us all to focus our ideas on maths and develop some clarity about what, how and when we teach maths;
- helping new teachers or those who feel insecure teaching maths to 'sort their maths out'.

'A scheme of work is the essential working document of classroom practice . . . [it will be] largely concerned with details of knowledge, skills, and processes to be taught . . . Both writing and subsequent reviewing of a scheme of work should involve all the staff who are concerned with its teaching.' (Non-Statutory Guidance, B 5.0.)

A scheme of work is *not* a blueprint for ever. We all go on changing and (hopefully!) improving what we do, so it is to be expected that we will want the scheme of work to be different almost as soon as we have written it.

As we work our way through what we thought was an adequate outline of how to cover two-dimensional shape with seven-year-olds, we get a brilliant idea, or something doesn't work as well as it did last year. So we add or cut out and that is good practice! The worrying scheme of work is the neat, unread and unmarked document in a file.

A new teacher might sit through all the maths INSET meetings about developing the scheme of work and know exactly what is in it, but by using it, and because of the rapid learning that goes on in the first year or two of teaching, he or she might find themselves able to put new meaning onto things in the scheme of work. Inevitably, he or she will want to change and develop as new ideas are found, or discoveries made about what works.

This highlights the need for yearly reviews of the school scheme of work, and one of the things that can go wrong with that is the time pressures in primary schools and the consequent difficulty of setting aside enough time to do reviews.

HOW THE SCHEME OF WORK RELATES TO THE NATIONAL CURRICULUM

The Programmes of Study will form the basis of the maths that is planned for in the scheme of work. Figure A5 shows the sort of process that the Non-Statutory Guidance sees us going through as we plan both on a long-term basis and for our daily work with children.

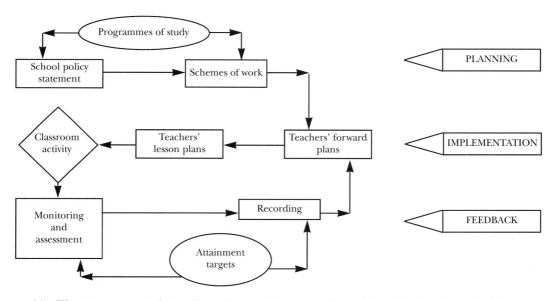

FIGURE A5 *The stages you might go through in writing your scheme of work (taken from the Non-Statutory Guidance).*

Explaining this process and enabling you to clarify what you are doing and why is one of the main aims of this book.

INTERPRETING THE PROGRAMMES OF STUDY

It is obvious as you read any of the curriculum documents of the British Isles that there are huge gaps in the Programmes of Study. Children don't suddenly develop a knowledge and working understanding of place value. It starts with games and counting in the reception class and develops through using calculators, Dienes apparatus, playing with number lines, doing games that involve adding and subtracting tens, doing grouping and exchanging activities and learning the role of different columns and bases in our number system. The statements in the curriculum cannot possibly represent the complexity of that learning process.

Similarly, we see gaps in any curriculum if we teach in the early years because, as we look at what a child needs to be able to do at about eight, we see that we lay the foundations for that at ages four, five and six. Probability is a good example here. Just because probability is not explicitly on the curriculum for a five- and six-year-old does not mean that we should not do any probability. It is important that children learn to use the language of probability and to understand likelihoods of events happening: Do you think it might rain today? Will Jani's mum have a boy or a girl? It is impossible for me to be six tomorrow if I am only five and half today. If I jump up in the air, is it certain that I will come down again? Is it harder to throw a six on a dice than all the other numbers? If I toss this coin a hundred times, what might happen?

All of those experiences need to be part of a young child's early mathematical experience in order for them to learn securely later on. If we take other examples, such as an early understanding of grouping that can lay the foundation for multiplication and division, or early experiences with cutting up play-dough jam tarts for the teddies that gets children thinking about fractions, we see that we need to look beyond our own Key Stage to see what we need to be doing that might not be explicitly mentioned in the curriculum.

Giving in-depth experiences

You need to cover the curriculum, but almost certainly you can't cover everything in as much depth as you might like. Giving children in-depth experiences of some of the curriculum is maybe a better experience than rushing through lots of content superficially.

Some attempt has been made with the English and Welsh curriculum to slim it down, and this helps, but you might think that there is still rather a lot to cover, so plan to do some bits in more detail than others – or, if your children become hooked on Logo, build on that and let them get deeper and deeper into it. The experience of a deep level of engagement with something they really love will help to develop those essential positive attitudes to maths and to learning in general.

Planning open-ended maths

(See also pages 33 – 9.)

One of the most worrying things that advisers and inspectors report as they go around schools is that there is still much less use of open-ended problems than they would like to see. During INSET, teachers often say to me that they simply don't have the time to do open-ended problems. Teaching the curriculum as it stands is itself so demanding that they must just teach specific tasks so that they can tick off having covered something on their plan.

This is an understandable position in some ways. We have to cover the curriculum. We have to convince the head and the governors that we have done so, and for most of us, life with thirty small children for five days a week is so demanding that if we didn't find at least some ways to make our life less complex, even more of us would suffer stress-related illness!

But – and this really is a huge 'but' – although using open-ended problems takes some getting used to, and it needs planning and a bit of confidence to deal with the likelihood that children will come up with things that we never thought of, open-ended problems can make our job very much easier.

USING OPEN-ENDED TASKS

Open-ended tasks make our job easier because:

- these kinds of tasks involve so many areas of maths that they enable us to cover the maths curriculum with far fewer tasks;
- they are the *only* way to cover the requirements for problem solving, investigating and communicating mathematics (using and applying maths in the English and Welsh curriculum);
- they involve the children in talking about maths and therefore the children learn more thoroughly and more securely;
- the children need to think about what they are doing, and in this way, they learn very much more than just filling in boxes or working out arithmetic examples;
- they build up children's confidence so children begin to see themselves as confident mathematicians and learn to tackle any problem given to them;
- children start to see the relevance of maths to their daily life because they are involved in measuring, calculating (usually in their own way) and using the skills that they see parents and teachers needing for *their* daily lives.

OPEN-ENDED TASKS CAN GO ON EVERY DAY

You might well have seen or heard of teachers who have done very long problem-solving tasks with their class, and this might well have put you off (see below). However, although this kind of complex task can be incredibly rewarding for the teacher and child, open-ended tasks don't need to be long.

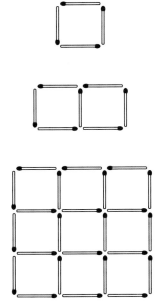

FIGURE A6 *A growing pattern made with matchsticks.*

IDEAS FOR QUICK OPEN-ENDED TASKS

- Could you build a kennel for Spot out of Multilink? Estimate how many Multilink you might need before you start. Then count how many you use. Make sure Spot can get through the door! (Using and applying, number, area, linear measures.)
- There are thirty-two people here today, and we need to get into equal-sized groups. How big could the groups be? (Using and applying, number – multiplication and division.)
- 'Fifteen' is the answer, so what could the question be? (Using and applying, number – could include fractions, decimals, multiplication, division.)
- Could you build a tall tower out of these cereal boxes? (Using and applying, properties of shapes, linear measures.)
- Could you make an open box out of this piece of A4 paper? Now could you make one twice as tall? (Using and applying, shape, measures.)
- I've started a pattern here with matchsticks (see figure A6). Can you continue it? Each shape gets bigger than the one before. (Using and applying, pattern.)

IDEAS THAT MIGHT NEED A BIT MORE TIME

- Let's make a time line that fits all round the room and that is big enough to show some of the details of the events in Victorian times. What would be a good length of line to use for a century? (Using and applying, linear measures, place value, proportions.)
- What is the tallest tower that you can make with fifty Lego bricks and that will support a 50g weight? (Using and applying, weighing, linear measures.)
- Let's plan a party for Friday afternoon. (Using and applying, capacity (how much drink do we need?), data handling (which kind of crisps is the favourite?), money (how much will it cost?).)
- Here are ten newspapers, a kilogram weight and a roll of sticky tape. Can you find a way of suspending the weight a metre from the floor just using the newspaper and tape? (Using and applying, measures.)

Those long and complex (but hugely enjoyable and profitable) problem-solving projects include:

- We're going to spend a week away on Anglesey in May. Let's plan what we need to take, what we could do and how much it will cost.
- Let's plan a sports day for the whole school.
- It would be good if we have a natural area and pond in our school grounds.

- Let's make the playground much more interesting.
- Could we have a youth club?
- Let's go out and have a picnic.

A teacher I know planned a sports day with his class. It took over half a term to plan and carry out, and it covered a huge amount of the maths curriculum. Not all children covered all the maths in the same amount of detail because the children worked in groups assigned to different tasks, but each child experienced some of all these areas of maths during class 'review' or 'carpet' times. Some teachers use this as a reason for not doing this kind of open-ended task – 'How will I know what each child has learnt?' – but we don't know what a child has learnt in any task. It is only when the children have to apply their knowledge in this kind of problem-solving situation that we can truly say that a child really understands something.

This was the maths that the children covered when they worked on their sports day:

- linear measurement (planning the sizes of the running tracks for the different age groups);
- measuring time (how long will it last, how long should we allow between races and how will we time the races so that we can start some school records?);
- planning and allocating space (they had long jump, high jump and throwing wellies);
- map making (they made a plan of where spectators sat);
- planning and delegating (they wanted a refreshment stall for cups of tea, sandwiches and cakes);
- money (we need to buy drinks and food for the refreshment stall; we want to make a profit, so how much do we charge for a cup of tea? How many tea bags will we need to buy, etc.?);
- decimals (if we buy the largest box of tea bags and the largest bottles of milk, what is the minimum we can sell a cup of tea at?);
- capacity (if every child in the school is allowed two drinks, how much drink will we need? If the tea urn holds two gallons, will that be enough for all the parents to have a cup of tea?).

There was other organising that also required mathematical thinking, for example they made certificates for every child that took part and medals for first, second and third place; they realised that they would need considerable adult help to actually run the afternoon, as they all wanted to take part, and they would need to borrow stop watches, as they wanted to time each race, and this required them to make telephone calls, write letters and use considerable life skills in convincing these adults to give up an afternoon for what they insisted was a good cause.

Encouraged by their teacher, they also started a data base of the times they took to run 100 metres, and they realised that they could then see if next year they were any quicker. This led to timing some much younger children and putting this on the data base, along with height jump measurements, so that these could be used by the younger children when they were nine and ten years old.

It was a huge project and greatly admired by the parents and governors. The children were enormously enthusiastic about it and clearly loved being mathematicians. Even the least able child (age nine) was working with seconds on a stop watch, happily working out the cost of a cup of tea and proud to announce, when all the bills had been paid, that they made £9.14 profit. As his teacher said, 'he would never have dealt with that as a problem in the maths text book – he wouldn't have known where to start'.

Your open-ended problems don't have to be this grand! But don't forget that they can be the way that your children might learn maths best – so build them into your scheme of work where you can! Remember that:

- they help to cover lots of different areas of maths at one time;
- they can be wonderfully cross curricular;
- they are the best way that I know to get children saying, 'please can I do some more maths?'

- What do I really enjoy teaching in maths?
- Can we agree that, if I do more of that this year and show that in my plans, then the next teacher will do more of something he or she really likes doing so that in the end the children get a balance?
- Can we decide what our scheme of work is for?
- Can we decide on a way of presenting it? (Loose-leaf, done on a word processor, lots of wide margins for notes? Can we all have a copy?, etc.)
- Can we select some formats that we could trial?
- Could we all have a go at one simple, open-ended problem in the next month and then report back on how we got on? (For examples of these, see pages 14 and 33 or look in *Bounce to it* or some of the other books in the resources list.)

You could consider this list as individuals and make notes to take to the next staff meeting.

CHAPTER 2

ROLES WITHIN THE SCHOOL

This section includes:
1 The collective roles
2 The role of the maths co-ordinator
3 The role of the head
4 The management of change

THE COLLECTIVE ROLES

From our experience, it is the pulling together – the sense of partnership – that seems to characterise successful schools. It would make sense to say that if that partnership is so crucial, then it would seem that everyone has a role to play in the successful putting together and implementation of the school scheme of work. The views and needs of the smallest child, the most awkward parent, the busiest governor and the teacher who insists that she is 'hopeless' at maths all need to be accommodated in what we end up with. We all have a role to play, but we don't all need to play the same role or have the same time commitment to the task.

Much of our role is about supporting each other. When we work in a successful, happy school, teachers often identify the reason why the school is so happy as something like 'everyone is very positive and supportive'. Again and again we hear that said in schools, and it is clear that you don't need to agree with everything someone does for you to support them.

When we work together doing INSET, conflicts and disagreements are bound to arise. It would be very worrying if they didn't! Conflict can be the beginning of change.

THE ROLE OF THE MATHS CO-ORDINATOR

The single most important role in terms of actually getting something done is that of the maths co-ordinator. The role is complex and incredibly difficult! We need to try to be:

- facilitator;
- supporter;
- initiator;
- inspirer;
- encourager;
- learner – not the 'I know everything approach'.

Who has defined the role of the maths co-ordinator?

You may well have a job description that defines your role, or be in the process of putting one together. (One example is given in appendix 1.) You can get useful guidance from the Cockcroft outline of roles (see below), but you might have to update it for your role in a post-National Curriculum era and for your particular school. (If you want to read more about the co-ordinator role, you might find some of the books in the resources section useful.)

The work of the co-ordinator seems to divide into three main areas.

1 How to manage the subject matter of the mathematics, advising colleagues and organising the writing of various documents for the school, such as policy documents and a scheme of work.
2 The physical care of resources (e.g. organising a 'maths cupboard' or some kind of system to make the best use of resources).
3 The much more difficult and ill-defined role of 'effectively supporting' (Campbell, 1985) colleagues and engaging them in meaningful in-service work and professional development.

It is this third aspect of the role of the co-ordinator that is often the focus of most of our problems. We might feel that we have very little experience to draw on for our role as in-service provider. That is why we have included some possible activities and guidance at the end of each chapter and why we need also to listen and learn from others in middle management, our head teacher and other advisers who have had experience in managing change.

The influential Cockcroft Report (1982) gave clear guidance on the role of 'mathematics co-ordinator' (354–8). According to this report, the role includes:

- preparing a scheme of work for the school in consultation with the head and the staff;

- providing guidance and support to other members of staff, by meetings and working alongside individual teachers;
- organising and buying teaching resources;
- monitoring mathematics throughout the school, including methods of assessment and record keeping;
- assisting with the diagnosis of children's learning difficulties;
- arranging school-based in-service training for members of staff;
- liaison with the schools from which children come and to which they go, and also with LEA advisory staff;
- keeping up to date with current developments in maths education;
- helping young teachers and colleagues who lack confidence in maths.

It is clear from this list that a co-ordinator emerges 'as a major figure in the national attempt to protect and renew the primary school curriculum' (Campbell, 1985, page 53).

The problems for the co-ordinator

It is possible that co-ordinators of 'fringe' subjects such as music or PE are less likely to be seen as threatening and need hardly influence anyone's self-esteem. However, with a subject such as maths that everyone is expected to teach well, we need to be aware of other's feelings and remember that some teachers worry about asking for help.

We also need to be particularly aware of potentially de-skilling some colleagues. They might have considerable strengths in teaching maths, but they might feel they need to keep quiet as they are not the co-ordinator. They could feel unable to offer their own expertise in the presence of the so-called 'expert'. For older teachers there has been a huge move from autonomy in their classroom that can encourage creativity, openness and discovery learning to a much more constrained system of accountability. This is difficult to deal with for some, and we need to be sensitive to that.

Although there are some clear expectations of the role of a maths co-ordinator from the Cockcroft Report, how to actually turn those expectations into real action is unclear. The role can be so complex that it is overwhelming and fraught with conflicts. Problematic aspects of this uncertain and overwhelming role can take the form of hostility expressed towards the co-ordinator (for a variety of reasons), a lack of clarity about roles and the structure of the power positions within the school (we might be struggling with an ill-defined role in the school), and the nature of a new role within a school in which the co-ordinator needs to gain experience. Our context could be one of such rapid change that morale is low and the immense time pressures have led to 'burn-out' amongst colleagues. We also inevitably have other problems, such as staff turnover, a new head, key people being on leave, etc.

Much of the role needs negotiating, and that isn't easy. For example, a 'monitoring' role in the school is very much harder to carry out than organising equipment. As one maths co-ordinator said to me, 'I can monitor what the young teachers are doing – they ask for help and expect me to oversee their work, but some teachers are much older than me and I can't say to them 'let me see what you are doing', can I? We have quite enough wars in our school without me starting more!'

So the maths co-ordinator role is delicate, demanding and difficult. As we plan our school-based INSET, we need to look at our constraints and problems and plan with those in mind. Some of the role needs to be done with others – you would be wise not just to write the scheme of work for the school by yourself, for example – but some things we can do on our own.

WHAT WE CAN DO ON OUR OWN
- Sort resources.
- Liaise with other schools.
- Keep up to date with current ideas, etc.

WHAT WE NEED THE HELP OF OTHERS TO DO
- Working groups.
- Planning.
- Making a maths games library, etc.

WHAT INVOLVES THE WHOLE SCHOOL
- Pulling together the writing of a maths policy document.
- Deciding on the final version of the scheme of work.
- Making final decisions about big maths purchases.
- Deciding what changes are wanted and needed in maths.
- Developing a reflective atmosphere, etc.

From my own research into my role, I drew out the following strategies for dealing with potential problems:

- It is likely that we will get on with some of our colleagues better than others, so concentrate on those colleagues who want to 'sort their maths out', as one of my colleagues put it. Of course we need to try to work with others who give you the cold shoulder; however, in a setting where time is limited, we will achieve much more if we work with those who want to take some action.

- Try to take a long-term view, e.g. think where can I get to in three years, rather than 'I've got to stop Eleanor doing that with her children before next Monday'.
- Recognise that much of what we do is about holding things in tension and that maybe some of these tensions cannot be resolved in the time available to us, with the resources or in our circumstances. For example, there could be a tension with using a scheme. Some teachers seem to need a scheme, but we might want to work without one, so the tension is learning how to use the scheme effectively but not slavishly.
- An obstructive head is incredibly difficult to deal with but is frighteningly common. My advice would be to be realistic, keep as low a profile as possible and be a bit subversive and do things on the quiet!
- It's not easy with a class teaching commitment to keep up with the latest work on maths, so you might want to consider belonging to an organisation that can help you to keep up to date. There are a few that focus on teaching maths, and the addresses for these are given in the resources list at the end of the book.

You will need to work on your strategies for attaining your identified goals. These might include:

- Finding ways to take the pressure off some colleagues.
- Finding ways to reassure colleagues.
- Working alongside colleagues (provided they agree) either in your or their classroom.
- Meeting with staff informally over cups of tea to discuss any topic that arises. (Don't underestimate the importance of informal discussions.)
- Looking out for ways to make colleagues' good practice explicit. Be generous with praise (after class assemblies, etc.) and help with things such as making a maths display. It is often those good relationships and straightforward friendships that can be the starting point for professional development.

THE ROLE OF THE HEAD

Whatever you read about effective schools, management issues, the success of the co-ordinator role, the success of INSET or anything else than can go on in a school, often the most important aspect of success turns on the way the head carries out his or her role.

However much teachers successfully get around a head who is less than supportive, in the end what we can offer the children and parents is severely constrained without the head's active and thoughtful support. Abdicating all responsibility to the co-ordinator is no good. We don't want the 'well, you just get on and do it and let me know when it is done' approach. We need a listening ear and thoughtful and constructive criticisms of our plans. Telling the co-ordinator to do the scheme of work and run INSET in the school but actually not enabling that to take place effectively (e.g. by still 'leading' the meeting when the co-ordinator is trying their best to be 'leader') is no good either because the co-ordinator is left with all of the responsibility to get something done but none of the authority to do it.

What we hope for is a collaboration. The head still carries the can, so they will obviously (usually) want to be involved at each stage and we must keep in touch with them. It is their school, and our plans, ideas, etc. must be aired. If your head is the 'I can only listen to you for four seconds' type, then write memos. Date them and keep copies. If they don't reply or don't read them, try to discuss it, but if you still get the cold shoulder, that is their problem and we need to feel our way forward sensitively.

If you are a head and you have a maths co-ordinator, you will want to discuss their role with them. You could use the job description outline (see appendix A) for that, or you could work with them on identifying what is going on in maths in school at the moment (see page 130 where there is a questionnaire that you could ask people to fill in, or fill in yourself via discussion with staff).

For a head, another very non-threatening way to initiate thinking about maths in the school is to do something about the resources in the school. This could be particularly successful if you are able to come up with some money for new equipment or at least some statement of the order of priority of spending the small amount of money that you have. On page 104, there is a list of questions that you could ask yourself or your staff to establish what you already have and what you need, and on page 107, there is a section on how you can evaluate published maths resources including schemes.

THE MANAGEMENT OF CHANGE

Many heads and some deputies have been on management training courses and know some of the pitfalls and complexities about change. This kind of training is harder to get if you are a co-ordinator, so it might be helpful for heads to share some of their insights into the successful management of people and ways of running successful INSET. Most heads will have some experience of running INSET and writing a scheme of work, and sharing these experiences can contribute enormously to the success of what a co-ordinator is trying to do.

Ten assumptions about change

This is what Michael Fullan (1982, page 92) suggests we take as our assumptions about change.

1   Do not assume that your version of what the change should be is the one that should or could be implemented. On the contrary, assume that one of the main purposes of the process of implementation is to **exchange your reality** of what should be through interaction with [...] others [...].

2   Assume that any significant innovation, if it is to result in change, requires individual implementers to work out their own meaning. Significant change involves a certain amount of ambiguity, ambivalence, and uncertainty for the individual about the meaning of change. Thus, effective implementation is a **process of clarification**.

3   Assume that conflict and disagreement are not only inevitable but fundamental to successful change [...].

4   Assume that people need pressure to change (even in a direction which they desire), but it will only be effective under conditions which allow them to react, to form their own position, to interact [...] to obtain technical assistance,etc. [...].

5   Assume that effective change takes time [...] Expect significant change to take a minimum of two or three years.

6   Do not assume that the reasons for lack of implementation is outright rejection of the values embodied in the change or hard-core resistance to all change. Assume that there are a number of possible reasons: value rejection, inadequate resources to support implementation, insufficient time elapsed.

7   Do not expect all or even most people or groups to change. The complexity of change is such that it is totally impossible to bring about widespread reform in any large social system. Progress occurs when we take steps (e.g. by following the assumptions listed here) which **increase** the number of people affected [...] Instead of being discouraged by all that

remains to be done, be encouraged by what has been accomplished [...].

8 Assume that you will need a plan which is based on the above assumptions and which addresses the factors known to affect implementation[...] Knowledge of the change process is essential.

9 Assume that no amount of knowledge will ever make it totally clear what action should be taken. Action decisions are a combination of valid knowledge, political considerations, on-the-spot decisions and intuition [...].

10 Assume that change is a frustrating, discouraging business. If all or some of the above assumptions cannot be made (a distinct possibility in some situations for some changes), do not expect significant change **as far as implementation is concerned.** [Bold the author's.]

Writing a scheme of work, implementing it, finding the resources to carry it out and evaluating what you are doing is definitely not easy! No change ever is. So be realistic in what you attempt to do.

WHOLE SCHOOL INVOLVEMENT

All the energy and enthusiasm from a maths co-ordinator, all the support from the head will all come to little unless the individual teachers become involved at an active and reflective level. One 'side' will always 'blame' the other for ineffectiveness, but actually implementing what we discuss in the staffroom is in the hands of the teacher.

It just isn't enough to say 'tell me what to do and I'll do it', though that is an amazingly common response from teachers to the co-ordinator. A good co-ordinator doesn't want to 'tell' teachers what to do. He or she wants comment, ideas, response and a sense of commitment – but equally a good co-ordinator needs to recognise that all classroom teachers have so much demanded of them that their top priority in terms of time and effort needs to be with the children. Having said that, though, working together on the scheme of work is crucial. We need to be realistic. Obviously staying at school for a two-hour meeting every week and things to do in class in-between times is going to be too much. Goals need to be realistic. Not everyone is ready and able to change their practice – and why should they? Maybe what they do already is pretty good! Change for the sake of change is unlikely to lead to successful innovation.

Negotiation is the key

There are some things that need to be addressed by all teachers. We need active and reflective involvement from everyone, but not necessarily a huge time commitment. We can:

- consider how the pressure can be taken off;
- consider giving non-contact time or working alongside each other in classrooms;
- consider thinking about what is on the school development plan for five minutes during our small group curriculum planning sessions with our close colleagues (or however you do your planning). How is this experience we are planning for the children fitting into the general development for the school?;
- consider working in small groups. See figure A7 for one possible working pattern.

FIGURE A7 *A possible working pattern for spreading the load of work.*

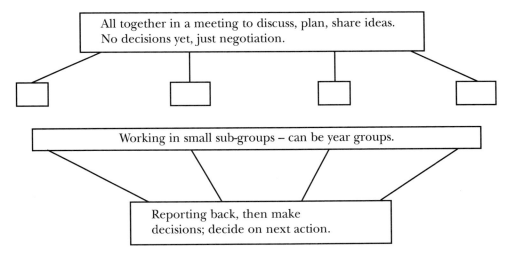

All together in a meeting to discuss, plan, share ideas. No decisions yet, just negotiation.

Working in small sub-groups – can be year groups.

Reporting back, then make decisions; decide on next action.

- What is my role?
- Am I making the best of it?
- What might help me in my role? Can we discuss that for a few minutes?
- How do we as a group deal with disagreements when they arise?
- Can we think how we can improve that?
- Do we need some ground rules for staff meetings?

You could use the 'action plan' in appendix 2 for individual responses to this.

FIGURE A8 *A typical INSET meeting.*

SECTION B

CREATING THE SCHEME OF WORK

This section is divided into four chapters:

CHAPTER 3
Defining the mathematical content

CHAPTER 4
Defining teaching and learning styles

CHAPTER 5
The order in which to teach the maths

CHAPTER 6
How long to spend on the maths

CHAPTER **3**

DEFINING THE MATHEMATICAL CONTENT

This chapter includes:
1   Using and applying maths
2   Number, shape, space and measures, data handling
3   Real-life mathematics
4   Open-ended activities
5   A balanced diet
6   Wonder and delight in maths
7   Maths from stories

USING AND APPLYING MATHS

However we plan our scheme of work, we need to make sure that our plans start with the Programmes of Study and that they cover all the required areas. As discussed above on page 12, we need also to fill the gaps in the curriculum (obviously no curriculum can fill in every step that a child might need to go through to learn a certain concept) and also we need to make sure that we look ahead to see what we need to do with our children to ensure that they have the experiences needed for later work, even where this is not mentioned in the curriculum. (Reception teachers don't usually think of themselves as teachers of fractions, but they do ask children to share out the play dough so that everyone has a fair share.)

Using and applying maths is a common feature of the various curricular documents in the British Isles, and the importance of this area of maths can hardly be stressed enough. This is the area which is about children applying what they do, learning to deal with real problems, learning to communicate what they know and talking about maths in a meaningful way. This maths focuses on the mathematical thinking processes that we use whenever we do 'real' maths – as opposed to the mindless kind of filling in the empty boxes or just working out the answer that some schemes seem to have to excess. It is these thinking processes that we want to develop in our children in order to help them to become confident and able mathematicians. This maths really does need to be central to our teaching.

NUMBER, SHAPE, SPACE AND MEASURES, DATA HANDLING

The content is similar in all the curricular documents, but there are some quirky bits to a casual observer. It can look as if young children don't do algebra now because it is often called 'pattern' and absorbed into 'number'. It can look as if Key Stage 1 children don't do data handling in England and Wales, but of course they do, it has just become a part of 'number'. As mentioned on page 12, it might look as if young children don't need to do any probability, but they do because they need to understand words like 'certain', 'maybe', 'likely', etc., and in most mathematically rich early years classrooms, those words and concepts are being explored as a part of daily life.

So, if we put all the rather unexpected bits into the categories that we might more traditionally think of them in, we end up with a list of mathematical content that looks rather like the lists that schools had in their schemes of work or 'tick lists' some years ago!

NUMBER
(You might want to include money here.)

* Counting, reading, writing and ordering numbers.
* Estimation and approximation.
* Place value (this can be a sensible place to teach money).
* Four operations and the relationships between them.
* Using a calculator.
* Fractions, decimals, percentages.

Number is so closely linked with data handling, measures and algebra that many people like to follow the curriculum and keep them here in their scheme of work, but others like to think of them separately.

PATTERN OR ALGEBRA
Some people now incorporate this into 'number'.

* Patterns, relationships, sequences.
* Making generalisations (often about patterns, relationships and sequences).

- Functions, formulae, equations.
- Graphs that express relationships.

MEASURES

Some people like to include money here.

- Length.
- Weight.
- Area.
- Capacity and volume.
- Time.
- Temperature.
- Angle.

SHAPE AND SPACE

- Two- and three-dimensional shape – symmetry, transformations.
- Angle – movement, rotation, direction.
- Using Logo and a floor robot.

DATA HANDLING

- Collecting, representing and interpreting data.
- Using data handling packages on the computer.
- Probability.

Don't forget the problem solving and investigations which enable children to apply their knowledge and skills.

REAL-LIFE MATHS

We can make 'real' maths central to that by putting activities based on using and applying maths into the scheme of work.

Sounds obvious doesn't it? But you would be surprised how often schemes of work hardly mention using and applying maths, or say that using and applying maths is there in everything. That might be true (it depends how open-ended your 'content'-based activities (number, shape, etc.) are), but the disadvantage of just saying it is there in everything is that some teachers get so carried away with covering the content that they fail to address the very specific processes involved when using and applying maths. Then, when they have to assess children's maths, they are confused and concerned that they don't know if the children can do what is required.

Make sure that your 'content' (number, etc.) activities are as open-ended as possible so that children develop their mathematical thinking through covering this content.

You might well be able to write a good scheme of work that really does integrate mathematical thinking processes throughout, so you might decide that you don't want to have a separate section for this area of maths. If you do that, it would be appropriate to integrate the 'statements' of the Programmes of study for using and applying maths actually into the scheme of work alongside the content ones.

So, if you want children to 'relate numerals and other mathematical symbols, e.g. $+$, $=$, to a range of situations', you could do the 'cookies in a jar' activity below. (It can be conkers in a flower pot or buttons in a yoghurt pot). This gives children experience with number bonds to ten (or less) and helps them to begin to understand about unknown numbers (an aspect of pattern in number and the beginning of algebra).

INSET   The activity is suitable for reception onwards; it might be a good idea for everyone to do it in their classroom one week and then share observations at a meeting. If reception children have been kept to numbers up to five, this is the moment to break away from that!

Each pair of children has ten cookies (made from play dough or use counters) and a big yoghurt pot. One child secretly puts some of the cookies in the jar and the rest are on the table. The second child is then allowed to look at the remaining cookies and they must work out how many cookies there are in the jar. (So if seven are left on the table, they are working out $7 + 3 = 10$ in a way that is $7 + ... = 10$, hence the use of this activity for early number pattern and algebra work.)

The activity can go on for some time with children trying to work out if they have found all the different ways they can of putting some cookies in the jar and some of them out.

You might follow this activity up with children discussing at 'review' or 'carpet' time, and you also might want children to draw what they did. It is a very interesting activity to ask children to record in their own way; figure B1 shows some examples of children's work.

When we do this kind of activity with five- and six-year-olds, we might expect them not to use standard notations (e.g. $-$, $+$ and $=$) very much (they might do, especially if they have had lots of experience with a calculator). However, what I have found so interesting about this activity is that when children of seven or eight do it, they often surprise their teacher by *not* using standard

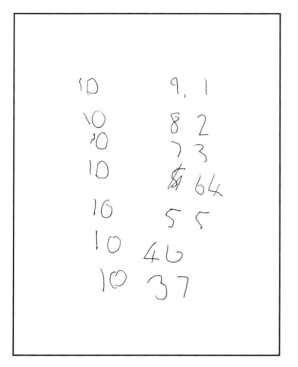

FIGURE B1a *Amira (6 years 5 months) was able to explain her work. She said, 'I can make ten in lots of ways.' When asked if there were more ways, she said, 'Yes, but I'm tired. You could have none in the jar and ten on the table.'*

FIGURE B1b *This was the first time that Tom (6 years 8 months) had recorded his work systematically.*

symbols for addition and subtraction, even although the teacher is convinced that they know and understand such notations. That is absolutely nothing to worry about because what we are trying to get at with this kind of activity is the kind of mental image that an individual child has. It is always helpful for us to try to find out the images children have in their minds, and if they are not using standard symbols, that's fine. It probably shows that they are thinking for themselves and not just doing what their teacher has told them to do.

So if you have this kind of activity in your scheme of work under number and pattern, you would also be including the using and applying aspect of maths.

CLOSED TASK	MODIFIED TASK
$2 + 6 - 3 =$	What numbers can you make from 2, 3 and 6?
$3 \times 5 =$	Make up some questions whose answer is 15.
(right triangle with sides 9, 4 and angle $x°$) Find the value of x	Investigate what the $\boxed{\sin}$ button on a calculator does.
Continue this sequence: 1, 2, 4	Discuss how the sequence 1, 2, 4 . . . . . might continue
(triangle) Find the area of this triangle	(triangle) Construct some triangles with the same area as this one.
What do we call a five-sided shape?	What shapes/configurations can you make with five lines?
Play a particular board-game	Design a board game for four people, using a die and counters
Draw the graphs of 1) $y = 3x + 5$ 2) $y = 2x - 4$ 3) $y = 6 - x$	Investigate the graphs of $y = ax + b$ for different values of a and b
Copy and complete this addition table	Investigate the possible ways of completing this table:

+	4	7
2		
6		

	3	4
	7	

FIGURE B2 *Closed to open activities (Non-Statutory Guidance).*

OPEN-ENDED ACTIVITIES

This has been discussed briefly on page 12, but it is so crucial that you might want to do an activity with your colleagues to focus on this point.

INSET ▶ If you have a copy of the Non-Statutory Guidance from the first version of the English and Welsh National Curriculum, you might remember a page of examples of 'closed' tasks and how we could make them open, (see figure B2 on page 33).

Once you have shown your colleagues this chart, you can look at your present resources or scheme and look for 'closed' tasks. Then try to find a way to make them open. Figure B3 shows some examples you might find.

FIGURE B3 *These tasks are all 'closed'. How would you change them into 'open' tasks?*

- 5 + ... = 10
- 3p + 4p + 3p =
- Copy this net of a tetrahedron onto card and make a tetrahedron.
- Measure these lines. Write 'line A is ...... cm long'.
- Which of these shapes are hexagons?
- Colour the set with more bricks.
- Find 20% of £200.
- 30 minutes is ...... of an hour.
- Write this time in figures: a quarter to nine.
- Copy these shapes, cut them out and fold them. Write, 'shape A has ...... axes of symmetry'.
- $147 \times 36 = $ ......
- Mr Smith buys 1200 bricks to build his garage. He uses 345 for the back wall. How many are left for the other walls?

Of course, not every task we give to children needs to be open-ended. Sometimes we really do need to know if a child can add up five and four, or find out if they know what a pentagon looks like; but mostly, if we use predominantly open-ended tasks in our classroom, we will know much more clearly whether the child *really* understands.

Using open-ended tasks is one of the most useful ways of assessing what children know. Because they have to think about and talk about what they are doing, their thinking is clear when you observe them working and listen to what they are saying. Their thinking is 'out there on the table', as Professor Richard Skemp used to say about children doing mathematical games.

A BALANCED DIET

If we are truly going to cover using and applying maths sufficiently in our scheme of work, we must show that we are giving children a wide range of different experiences. A balanced diet means doing:

- practical work;
- mental maths;
- maths games;
- computer and calculator work;
- some short and some extended pieces of work;
- other aspects of maths (this is discussed in more detail on page 42).

If you don't plan exactly how you will do that, you almost certainly will not be covering some of the required curriculum, and the maths about 'how' our children (and we) do maths could get missed out. But it is those thinking processes that are so important if we or our children are to feel confident with maths.

For example, while doing the cookies in a jar activity above, some children would probably be showing evidence of thinking systematically.

In figure B4, Tamsin (aged five) clearly showed that she was thinking of five and zero, four and one, three and two, and two and three. She was also able to record systematically. When asked by her teacher if she thought there were any other ways, she said, 'You could have none in and five out and that's different from five in and none out'.

FIGURE B4 *Tamsin's recording of 'cookies in a jar'.*

Mathematical processes

There are a huge number of mathematical thinking skills and processes, and by giving our children a wide range of different kinds of activities, we can be more sure that they are covering the range of mathematical thinking that they need to become confident mathematicians.

INSET   It is not sufficient to give colleagues a list of maths processes to discuss. They are, by their nature, things that can only be known in an experiential kind of way. This can be done by observing children closely or (and better) it can be addressed by doing some kind of open-ended task as a staff working in small groups.

Here are some starting points. It would be best to choose just one of these to start with.

1 CANS OF BEANS 1

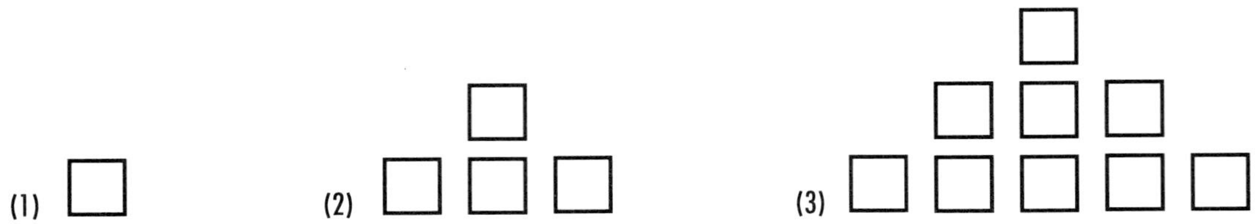

FIGURE B5 *Cans of beans 1.*

This is a terrific activity (see figure B5) because one of the patterns it generates is that the rows going downwards are the odd numbers (1, 3, 5, etc.) and another pattern from the totals of cans in each 'set' is the sequence of squared numbers (1, 4, 9, 16, 25, 36, etc.). It is a good idea to use Unifix cubes or similar to do the activity and to provide plain and squared paper.

If you ask why the squared numbers are found, it can stump some people, and this can reveal the belief that maths is so mysterious and so inexplicable that maybe there is no reason why the squared numbers appear. In fact, you can actually make squares if you turn a part of the 'stack of beans' round, as in figure B6.

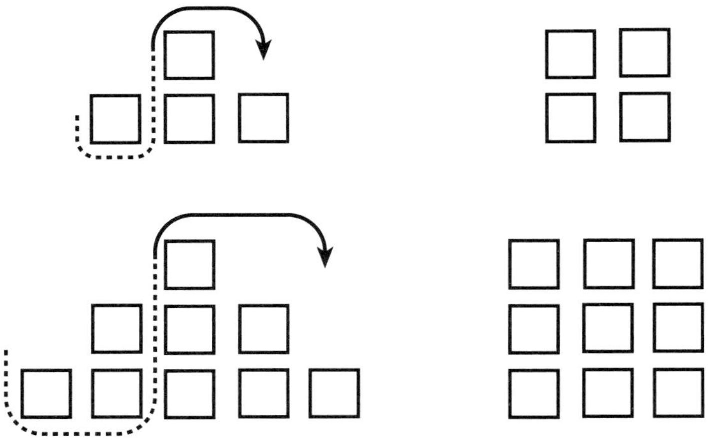

FIGURE B6 *When part of the pattern is moved around, it can be seen how the squared numbers are generated.*

It gives the squared numbers because there are the squares! This can be a real 'penny dropping' moment for some.

(Someone usually says, 'that this isn't how cans of beans are stacked in the supermarket' and so you might want to call it something else.)

2 CANS OF BEANS 2

Figure B7 shows the pattern of how cans actually are stacked in the supermarket. It generates the triangular numbers (1, 3, 6, 10) and these are made in an interesting way.

Pattern 1 has 1 can.
Pattern 2 has 1 + 2.
Pattern 3 has 1 + 2 + 3.
Pattern 4 has 1 + 2 + 3 + 4.
And so on.

FIGURE B7 *Cans of beans 2.*

It is good to provide dotty triangular grid paper for this activity, and you might want cubes to work with as well.

As with any growing pattern, it is helpful to suggest that people try to think of a way of working out (without making it or drawing it) what the thirty-first, or hundredth pattern might look like. They are thus getting some experience of what learning to generalise is about. They do a few special examples (so they are specialising) and then they can say something like 'oh, I see. The hundredth pattern would have...' That is generalising and it is an important mathematical thinking process. Whenever we teach maths, we hope that, by focusing on a few special cases, children will draw some general conclusions.

3 PAINTED CUBES

You need lots of interlocking cubes to do this, squared paper and at least half an hour. (The cans of beans investigations are much quicker). The painted cubes' exercise is great for an INSET day, and it is good to get colleagues to work in small groups and get each small group to report their findings to the other groups.

You need to encourage colleagues to think how many painted faces there would be for any size of cube, so can they work out, without making it, how many faces on a six-by-six cube would have three painted faces?.

Imagine a cube made with interlocking cubes that has sides three cubes long.

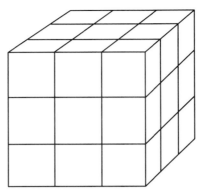

This is dropped into paint! (You can probably imagine that bit!)

Explore how many of the interlocking cubes have
• no faces painted
• one face painted
• two faces painted
• three faces painted.

What if the cube was a different size, e.g. 4 by 4, or 5 by 5?

FIGURE B8 *Painted cubes*

If you have a number of very nervous mathematicians in your group, you will need to be very encouraging and to sit them with people who will not put them down. The investigation needs some persistence, but the pay-off can be enormous as colleagues find ways of communicating what they are thinking. However, if your school has people in positions of authority who are likely to be openly negative as you work together, it is best to avoid this, or any other quite demanding activity. Go for a less enthralling and but much less complex activity like one of the 'cans of beans'.

Whatever activity you do together, it is good to ask colleagues to try to identify what and how they are thinking as they work together. It can help if the INSET provider goes around the groups listening to colleagues working and identifies the mathematical thinking and writes it up on a flip chart or overhead projector.

The kinds of thinking processes that you might be able to identify include those in figure B9.

You can do either of the 'cans of beans' investigations with children from about seven, and 'painted cubes' from about eight or nine. You could put some of the maths thinking processes up on the classroom maths board maybe as 'think clouds'. For example, doing either of the 'cans of beans' activity clearly shows

---

SOME MATHEMATICAL THINKING PROCESSES

- choosing appropriate equipment
- working systematically
- stopping and reviewing progress
- making a difficult task more simple
- presenting results clearly
- classifying
- deciding what information is relevant
- finding ways to record results

- finding ways round difficulties
- estimating
- thinking ahead
- talking about your work
- generalising
- creating new problems to help solve old ones
- explaining how or why something works
- explaining results to others
- putting things in order
- using trial and error

- predicting
- talking about work with others
- seeing relationships between things
- working in other examples
- trying different strategies
- looking for patterns
- interpreting information appropriately
- estimating

FIGURE B9 *The list of thinking processes is very long; you can add to it as you work together.*

the process of looking for pattern and predicting what is going to happen among many others that will emerge as you do the activity.

WONDER AND DELIGHT IN MATHS

If you do any of the investigations in the previous section, particularly if you do 'painted cubes' and it goes well, you can get the kind of reaction that shows the enormous power of maths to amaze us. A very important aspect of doing using and applying maths (some kind of problem solving and investigation) is that it helps children to enter into the wonder and delight of maths, or the aesthetic appreciation of patterns.

'As a complement to work (on) mathematics as a tool for everyday life, pupils should also have opportunities to explore and appreciate the structure of mathematics itself ... Mathematics is not only taught because it is useful. It should also be a source of delight and wonder, offering pupils intellectual excitement and as appreciation of its essential creativity'. (Non-Statutory Guidance, A. 2.5.)

MATHS FROM STORIES

One school teaches all their children (ages three to eight) from stories. Their scheme of work is based on the five areas of maths above plus using and applying maths, and they made their scheme of work by writing down all the 'statements' (e.g. 'addition facts to 20' and 'number bonds to 10') from the curriculum (plus the missing bits that they knew their young children would need to cover). They then had a piece of A4 paper for each story book and they put each 'statement' on at least one story book. So the counting books they used had the 'statements' for 'number bonds to 10'.

They use many different story books to do this and also have a list of supplementary books which they could use, but which they did not plan for specifically in their scheme of work. Figure B10 shows one of their pages based on the three bears. (If you want to consider this as a part of your scheme of work, see the resources list at the back for books to help you to do this.)

- Work with a colleague on the maths that you want your children to do this week and turn closed tasks into open ones.
- Think about how you like to plan your maths in terms of the areas of maths. Do you find it helpful to think of the five areas listed above or do you prefer to think of fewer areas and merge some, e.g. number and algebra?
- Brainstorm the maths 'topics' that you like to teach, e.g. money, two-dimensional shape, tessellations, co-ordinates, etc. (Make sure you cover everything in your curriculum.) This could form the basis of the groups of maths for your scheme of work. Don't forget using and applying!

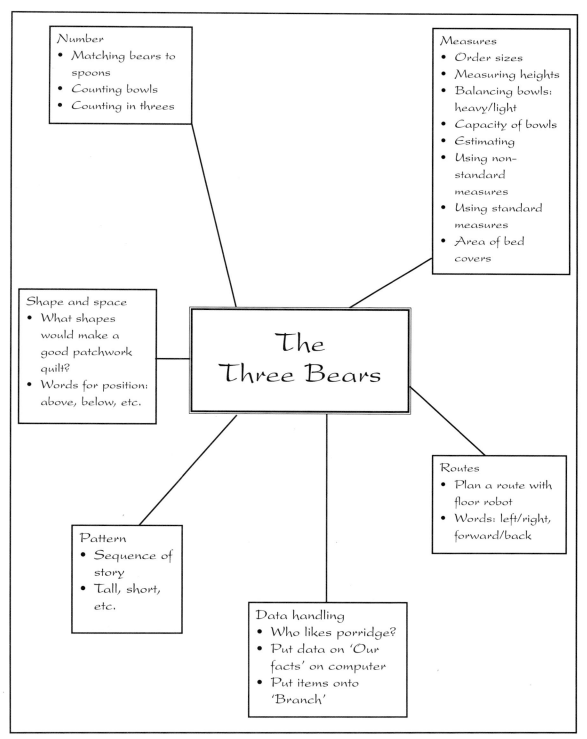

Number
- Matching bears to spoons
- Counting bowls
- Counting in threes

Measures
- Order sizes
- Measuring heights
- Balancing bowls: heavy/light
- Capacity of bowls
- Estimating
- Using non-standard measures
- Using standard measures
- Area of bed covers

Shape and space
- What shapes would make a good patchwork quilt?
- Words for position: above, below, etc.

The Three Bears

Routes
- Plan a route with floor robot
- Words: left/right, forward/back

Pattern
- Sequence of story
- Tall, short, etc.

Data handling
- Who likes porridge?
- Put data on 'Our facts' on computer
- Put items onto 'Branch'

FIGURE B10 *Scheme of work, format 3.*

CHAPTER 4

DEFINING TEACHING AND LEARNING STYLES

This chapter includes:
1   The balanced diet
2   Which comes first: the skill or the problem?

THE BALANCED DIET

There are many different aspects of maths (practical maths, problem solving, games, consolidation, etc.), and unless we address all these different kinds of ways of teaching maths, we are unlikely to fulfil our statutory requirement to teach the applications of maths and problem solving, etc. as well as the 'content' areas of number and shape, etc.

The need for these different areas of maths came to the attention of teachers after the Cockcroft Report (paragraph 243), and it is now widely accepted as a crucial part of our maths teaching.

This is often referred to as 'giving children a balanced diet of maths'. That's quite a good way of looking at it because, for many of us, we realise that we were brought up on a rather monotonous diet of mental arithmetic every morning (where we wrote down the answer and had them marked out of twenty), arithmetic from a textbook every day, 'problems' (usually work sums) about twice a week, such things as:

'From a roll of curtain, 638 pieces, each 1 yard long, were cut. Find in furs., chs., yds., the length of the role.'

and

'60 gallons of milk were delivered to the dairyman. In the forenoon, he sold 37 gals. 1 qt. 1 pt. How much remained to be sold in the afternoon?'

(Both quotes from 'Sure Foundation Arithmetic', book 3, Lockhart and Young, 1941)

and arithmetic tests every Friday, Riveting stuff! This all went on into secondary school, except that we then had occasional geometry and algebra thrown in. Most of this work was taught by the teacher doing two or three examples on the board and then we were told to open the text book on page 26 and there would be thousands and thousands of sums! (At least so it seemed to me.) If you couldn't understand what to do (and I usually couldn't) then you asked for help, and in their great wisdom, most of my teachers would then proceed to do another example on the board! We were taught 'how' to do things. That was, so it seemed, all that mattered. You tried to do exactly what the teacher told you. You turned it upside-down and multiplied it, you added up the units and carried a one and borrowed a ten and all that other stuff we were taught.

Many of us found it complete gobbledegook and when we asked 'why' we did it like that, or asked what it meant, we were told that it didn't matter, this is how you do it and that is what matters if you are to get your ticks.

Teachers used one main teaching strategy – exposition. It is not that there is anything wrong with this, but it needs to be balanced with other strategies. 'Chalk and talk' can only take children so far. They need practical experience as well because this is an important part of children developing their language and their conceptual understanding. The problem with some exposition is that it is far too abstract, and children don't have any 'pegs' to hang the information on. It goes in one ear and out the other all too easily, as we all know! But that is not to say that children cannot engage in discussing quite abstract ideas. They can and they do, and if you want to develop this in your children, SAPERE (see resources list) have some very useful materials for cross-curricular work based on developing children's thinking.

We must do some 'teaching'!

I used to teach in a school where every child did about an hour of maths every day, on their own, from the scheme. Each child worked through every page in the scheme at their own rate, and then the teacher marked the books about once a week. The problem was that, over the term that I worked in this school, I came to see that very few of the children actually worked out the maths for themselves. They had a variety of strategies to get the right answer, including things like sitting next to someone who had done the page and asking them what to do, going from one teacher to the other (it was a two-teacher unit with sixty children) and gradually wheedling the answer out of them, getting the answer book out of the teacher's drawer, and so on. Very few children over the three months that I observed them actually worked out the maths.

When I reflected on this and when I observed this kind of teaching of maths in other classrooms, what is so obviously lacking is any 'teaching'! Teacher input for a child that is 'stuck' is almost always focused on getting the children to complete the page. This is 'how' you do it and this is what you need to do. The child completes the page (eventually), but there is little evidence that the child has grasped the concept.

So although most of us had far too much 'exposition' when we were at school, it is no better in today's schools where there is almost no exposition at all. Children are just as lost under both systems.

The planning sheet shown in figure B11 or some adaptation of it to suit your school (see figure D18 on page 157) might be useful to you to start thinking about planning a more 'balanced diet' for your class. You can change the sheet to relate to what you want to do in your school, for example you can have more types of maths in the left-hand column, e.g. you might all want to focus on always getting some short and some extended tasks, or in nursery and reception you might want to use the left-hand column for different aspects of play. This planning sheet (in figure B11) needs to be quite large to get the boxes large enough to get all the detail of tasks and resource references on clearly.

You can use this type of sheet as a basis for your scheme of work in that you can put the area of maths at the top, (place value, addition, two-dimensional shape, etc.) and then list the statements and match these next to the activities in each box (as far as you are able to do that. I find it really difficult to think of activities for all the boxes for all areas of maths!). What matters is that, over the term, your children need a balanced diet, so if you don't have a computer programme to teach algebra, that really doesn't matter because the computer is probably put to better use for data handling and Logo anyway.

CONSOLIDATION

This involves children going over an area of maths, or a skill, or some knowledge until they are confident with it and can do it 'perfectly'. For many of us, this is what we did endlessly as children! We must do it but only when it is needed.

Some colleagues might well believe that consolidation needs to be done almost every day, and others might not want to do that.

PLANNING SHEET

TOPIC ....   TEACHER ....

TERM ....   AGE GROUP ....

relevant practical	
mental maths	
exposition	
computer/floor robot	
calculator	
aesthetic/creative	
design/technology	
group work	
individual	
discussion	
games	
problem solving	
investigation	
consolidation	
other	

FIGURE B11 *Planning sheet.*

You might want to talk about your school policy on the issue of consolidation.

MENTAL MATHS

Consolidation can be done very effectively through mental maths (where children don't write anything down). Consider this list of mental maths for seven-year-olds who have just finished a maths topic on place value and have been working on capacity and volume in the water and sand tray.

- Twenty-seven is made up of two tens and seven units, so how could you make up twenty-three? Thirty-one? Forty-five? Sixty-seven? Nineteen?
- How many tens are there in 100?
- Can you remember how many centilitres make a litre?
- Can you remember how much that litre of water weighed when Tracy weighed it in the red jug?
- Can you write 101 in the air? (Many children want to write 1001.)
- (Holding up two differently shaped bottles.) Can you estimate how many yoghurt pots full of water these might hold?
- Do you remember when we measured how much this wide bowl held and this tall measuring cylinder held? You were surprised about something.
- Let's count in tens up to 100/500.
- Let's count in hundreds up to 1000/2000.
- Who can remember how to write a million?

In just a few minutes, these children will have had considerable consolidation of the work that they have just done and will be reminded about the capacity work that they are doing. The 'facts', such as 1000 millilitres make a litre, might be re-capped every day that week and then gone over maybe six or seven times during the rest of the term, probably with the teacher demonstrating that 1000, 100 and ten are useful numbers when it comes to using metric measures. (This is definitely the moment to tell them how lucky they are not to be doing rods, poles and perches and not to have twenty shillings in a pound and 240 pennies and however many half crowns it was!)

Of course, mental maths can be used as a starting point of a topic and not just for consolidation. I think that every class, from reception right up into the secondary school, would benefit from at least five minutes of mental maths every day. (This includes singing maths rhymes for younger children.)

Mental maths is vital to children developing their own methods of calculating, which are very important considering that

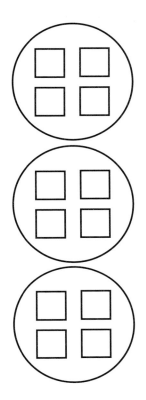

FIGURE B12 *Four lots of three.*

most of them will use calculators a great deal in their lives and for developing their mental images. This is a crucial aspect of our scheme of work.

EXPOSITION

Exposition can be:

- explaining the rules of a new maths game and explaining to the children what they will learn from playing the game. It is perfectly possible to show the whole class a maths game together sitting on the carpet. Then some of the children can go off and play the game. Then some of those children show some other children how to play it and so on. You can ask the children to play it on the carpet the next day in a sharing session just to remind them what to do and to check that they are playing it appropriately. Then another day you can teach a new game in the same way, so that you are building up a collection of useful activities in your class that can be 'teacher independent' and yet still a worthwhile activity that is a part of the curriculum;
- a 'review' or sharing session when different children explain their different ways of tackling the same thing and the teacher draws it all together, e.g. 'Shazad worked out fourteen lots of sixteen by putting out seven rows of sixteen and then...';
- teaching things such as four lots of three means four sets of Unifix set out like this in groups of three (see figure B12).

NB These examples used apparatus and not just words. So although it is 'exposition', it is not just 'chalk and talk'. The children have some physical object that they can relate to.

GAMES

There are a number of advantages in using games frequently in our classroom.

- They can help with consolidation of concepts. ('This knowledge [of number facts] was most successfully acquired where the work was practical, especially when number games were played.' OFSTED, 1993)
- They can be teacher independent.
- They can emphasise necessary language.
- They prompt child/child discussion as well as child/teacher discussion.
- They can encourage children to learn to make their own decisions.
- They can keep the child in the learning situation long enough for the concept to be experienced in some depth.

- They can help to develop positive attitudes to maths.
- They can be open-ended and prompt children to think divergently and creatively.
- They can provide a meaningful context for mathematical work and so children learn more securely.

COMPUTER, CALCULATORS AND FLOOR ROBOT

The importance of new technology to children can hardly be over emphasised. This is discussed in greater detail on pages 96 – 103, but when we are planning our activities for our scheme of work, new technology needs to be given a very high priority.

AESTHETIC/CREATIVE

It can sometimes be the aesthetic aspects that can switch a child onto maths. It is vital for children to have the time and space to develop their creativity, and doing that with compasses, rulers, two-dimensional shapes and grid papers, for example, can be not only a source of constructive kinds of teacher-independent activities, but so fundamentaly satisfying that children will want to take their work home to finish and share with parents. Any kind of pattern, design or exploration of a shape can form the basis for this kind of work, e.g. Islamic designs, Chinese tangrams or making patterns using just circles or parallel lines and circles. Children's work can be displayed by them or put in a class book, and it can be influential in developing awareness of shape and space.

DESIGN AND TECHNOLOGY

There are many mathematical aspects of technology and we can make those links explicit in our plans.

RELEVANT PRACTICAL

Many of us have experienced children apparently 'playing' with the Unifix and doing 'practical' maths in a way that makes the chimp's tea-party at the zoo look organised. It isn't just enough to give the children any old thing to do, or to put the apparatus on the desk and hope they will use it. It must be relevant, have a clear purpose and preferably be discussed by the children at 'review' time.

I have met teachers who are totally convinced that, once the children are seven, they don't need apparatus. They used to say, 'you're not having any of that baby stuff now. You're in the juniors now and you do it in your head'. This comes from an assumption that apparatus helps children form basic concepts only. Naturally enough, the children learn that using any kind of apparatus is 'babyish'. Some strategies for overcoming children's misconceptions are:

- You could try using the 'secondary' Multilink that is white, grey and black and used for GCSE. I know some eleven-year-olds who think it is 'cool'!
- You could try insisting that everyone uses apparatus and praising children who use it well.
- You could make it essential to use it to do the tasks, e.g. games.
- Make a big thing of building up the maths resources in the classrooms with older children. If you can possibly introduce things that they may not have seen before, that is a help.
- Hold a parents' workshop where they need to use cubes, etc. (e.g. do a probability workshop or one where they have to build up a pattern like the 'cans of beans' on pages 36 and 37). This might show how crucial some kind of physical manipulation in maths can be.

GROUP WORK/INDIVIDUAL WORK

Planning how we group children for our activities is so important that some schools actually put the grouping onto their scheme of work. You might not want to do that, but it then becomes a task that needs to be addressed in teacher's forward plans. Some tasks are best done in groups (e.g. problem solving, games) and some are best done individually (e.g. consolidation).

DISCUSSION

There is a need to plan for child/child and child/teacher discussion of different types, e.g. a group of children reporting on their work to the rest of the class, individuals explaining their methods of calculating to the rest of the class, games that require the child to say what they have done, etc. (There is a section in the resources list of books that are about the role of language in maths.)

PROBLEM SOLVING AND INVESTIGATIONS

Teachers often ask the difference between problem solving and investigations (see figure B13).

Problems and investigations are those open-ended types of tasks discussed earlier on page 12. These are so crucial to planning a balanced and adequate scheme of work that they appear frequently throughout this book (see pages 38 – 40 and pages 100 and 102), and there are suggested resources for starting points at the end of the book.

The activity on page 34 about making closed tasks open can

Problem solving	Investigations
• Usually real life	• Usually pure maths
• There is a definite answer, but no set route to that answer. For example, children might have many ideas about how they could make the playground more interesting or how much the trip is going to cost, but in the end, they must make decisions for what they actually put on the playground and how much they will charge for the trip.	• There are many different ways of exploring it. There might be an answer, but often there isn't, or there are so many possible outcomes that only some of them can be explored, for example the cans of beans or number chains.
• Some definite thing must happen. The cost of the trip must be covered and the playground needs to be more interesting.	• The outcomes can be hugely varied and they are not necessarily 'right' or 'wrong'. The outcomes can often be developed by asking a 'what if . . . ?' question, e.g. what if you did it with much larger numbers or if you did it on triangular paper instead of squared paper?
• You could see problem solving as like a whirlpool. It starts very large, but gradually focuses on one point.	• You could see investigations like a pebble being thrown in a pond. It starts with one small thing and the ripples go on getting bigger and bigger.

FIGURE B13 *People sometimes use the words 'problem solving' and 'investigation' to mean the same thing. In this book, I have given them both a specific meaning.*

contribute a lot to colleagues feeling more confident about using this kind of task, and if you have not already done so, some INSET around actually doing some open-ended work can be helpful (see page 40). Open activities make excellent starting points for the whole class, and this can help enormously with classroom organisation.

WHICH COMES FIRST, THE SKILL OR THE PROBLEM?

When we attempt to give children a balanced diet in this way, using open-ended tasks as a basis for much of the maths, we begin to see that, as children encounter 'real' problems (such as seven-year-olds trying to work out the length of their stride by measuring how far they go using ten strides, then using calculators to work out the length of one), they encounter situations where they discover that they need a new skill. The seven-year-olds investigating the length of their stride had not really noticed the 'little dot' in 35.7, and that was the moment for their teacher to introduce them to decimals. Some eight-year-olds measuring the rainfall each day wanted to know what 'average' meant. The children planning the sports day described on page 00 had never thought of whether 12.39 seconds was larger or smaller than 12.71 seconds, and the idea that you might split up a second into hundredths or thousandths was a fascinating way for them to explore new knowledge about place value and decimals.

People will probably always argue about whether we should teach the skill first and then get children to apply it in a problem-solving situation, but you need to have some sense of how you will address this problem in your scheme of work.

If you use the kind of scheme that has 'sums' on one page and then you turn over and there is a page of 'problems' on the next (that use the same rule, so a page of subtractions is followed by a page of subtraction 'problems' – as one child told me, that was how he knew that he needed to do subtraction!), you will implicitly be following the rule that the skill comes first followed by the application of it.

The skill first

DISADVANTAGES
* Slower children rarely get beyond the skill learning pages.
* Because the skill learning pages often do not use 'real' maths (by which I mean relevant and that makes 'human sense' (Donaldson, 1978) to the child), the meaning of the whole thing is not clear to the child, and they begin to get the sense that maths is not about anything that they can relate to at all.
* It is very difficult for us to assess what children are thinking and learning other than whether they have got their maths right or wrong.
* It is very difficult for us to see where the children need to go next in their maths, as we rarely take them to the edge of what they know and reveal what they need to know next. For some

children, this lack of challenge in maths makes them bored and can turn them off maths for the rest of their lives.

- Until children are doing some kind of problem where they have to apply their maths, it is incredibly difficult to assess if they really know how to do it. They might know it today, but what about next Tuesday or next year in the next class?

ADVANTAGES

- Children never stray into areas of maths that require them and the teacher to think for themselves. This makes it all very safe! But that makes it very dangerous, in terms of child boredom levels!
- You can have a list of what you know the children have covered in a term or year – and you can do the same next year and the next.
- You can plan well ahead and never need to change your plans.

The application first

Starting with skill using (so using problems as the starting point on a topic in maths, or maybe using some kind of investigation to teach some number work) would be a very different style of teaching. You could, for example, use the costing of a trip to teach seven-year-olds about money, or you could start a topic on subtraction with the investigation shown in figure B14.

If you start with this kind of activity, you can teach subtraction up to whatever numbers your children choose. So as children do these kinds of tasks, they will often come upon skills that they need to use and have forgotten, or new skills that they will need to be taught.

A group of children asked to find out the tallest thing in the school grounds may need help with using a clinometer or tape measure, and you might decide to encourage them to draw things to scale on squared paper so that they can show everyone what they have done and prove that the tree is taller than the telephone post. That might reveal that they don't know how to represent 13 metres on centimetre-squared paper and have no idea how to use the protractor to draw an angle of 26 degrees.

DISADVANTAGES

- Children need help in groups to be shown skills that they need (such as allowing for the 'dead space' at the start of a ruler), and this needs to have teacher time.
- You can never be quite sure what is going to happen when you pose a problem.

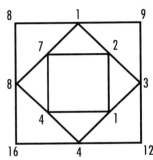

FIGURE B14 *Draw a square and put numbers at each corner. Find the difference between the numbers and put this half way along the line, creating another square inside. Go on doing this for as long as you can.*

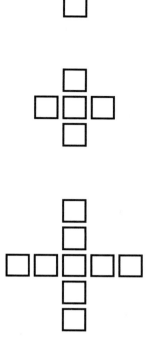

FIGURE B15 *This is a growing pattern called 'crosses' and is suitable for use with children from age five.*

- There are few 'right answers', and so children can become insecure if they are used to maths being rewarded with a tick; they sometimes need considerable help to be weaned off the idea that everything about maths is either right or wrong. This can take several weeks, and the more 'insecure' and 'tick driven' children take much longer. Lee (aged nine) said to me, 'I used to be good at maths, but now I have to think'. It took the whole term to get that amazing statement sorted out and get him back to feeling he was good at maths.
- If children are out and about in school, it can sometimes create problems of supervision and class management.

ADVANTAGES
- Children get used to the idea that maths involves them in their own thinking and that they need to be inventive.
- Children cover a huge number of the curriculum requirements just using one problem, e.g. the investigation of increasing sizes of crosses in figure B15 involves children in counting in fours (so it is good to do when you are teaching the four times table), it involves the use of numbers as far as children are prepared to go (one group of six-year-olds worked out successfully that the thousandth pattern would have 4001 cubes in it) and it involves the idea of generality. (They didn't make the thousandth pattern, they said it would be four lots of a thousand plus the one in the middle.)

If we think of giving children a balanced diet of using the skills they already have and acquiring new skills, we might think of our maths teaching as being a part of this cycle (see figure B16).

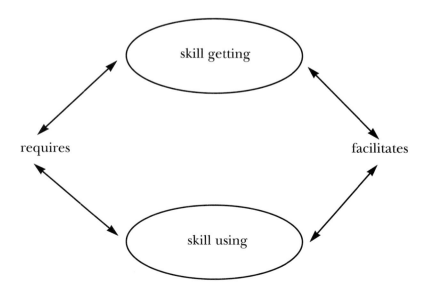

FIGURE B16 *The skill-getting and skill-using cycle (adapted from Open University, 1982).*

So a child using the skills of working out divisions and fractions in dividing up nine biscuits between five children might not know what to call the little 'bits', so the teacher can teach that. I believe that when children are taught something that they need, they seem to learn it quickly because they then use the skill and see the sense in it. Learning the name of a 'fraction' e.g. a sixth, could be pretty meaningless in abstract form.

It works the other way, too. So if a child is being taught a specific thing, perhaps how to use a ruler, it will be much more meaningful if the skill getting is put in a context that the child can relate to. A page of lines in a scheme to measure doesn't have much meaning, but measuring 70 cm of thread to do a three-point stitch on a home-made book does. Both activities can potentially tell you if a child can use a ruler appropriately.

> You might want to consider reproducing some of the statements in figure B17 for discussing during a staff meeting. The issue of 'standards' in education and things like teaching one set way of doing calculations is bound to arise in discussion. These statements are reproduced from McIntosh (1981), and the whole chapter makes illuminating reading for teachers and parents. You could, for example, reproduce some of the statements and see when colleagues and parents think they were written.

WHEN WILL THEY EVER LEARN?

1 'I must confess to some surprise at the extremely poor results in arithmetic.' (Intellectual Arithmetic, by a teacher of youth, page IV, 1840)

2 'In these standards [I and II – seven- and nine-year olds], at least, far too much time is given to the mechanical part of the subject. The results of this unintelligent teaching shows itself in the inability of the upper standards to solve simple problems.' (Board of Education, Special Reports on Educational Subjects, volume 26, page 16, 1912)

3 'There is a prevailing opinion that the London elementary school children of today are slower and less accurate in computation than they were ten years ago. I have searched for evidence in support of this contention, but have failed to find it. I am, therefore, inclined to relegate the belief to that group of opinions which have reference to the annual deterioration of Academy pictures, the increasing degeneracy of each new generation of men and other palpable fictions. But even if there has been a slight loss of accuracy, there has been a great gain of intelligence [and that is] incomparably more valuable than facility in calculation.' (HMI, 1895)

4 'If a child be requested to divide a number of apples among a certain number of persons, he will contrive a way to do it, and will tell how many each must have. The method which children take to do these things, though always correct, is not always the most expeditious. To succeed, it is necessary rather to furnish occasions for them to exercise their own skill in performing examples rather than to give them rules. They should be allowed to pursue their own method first, and then should be made to observe and explain it; and if it were not the best, some improvement should be suggested.' (HMI in Kent and Sussex, 1975)

FIGURE B17 *When will they ever learn? – thought-provoking statements for discussion from Floyd, 1981.*

THE ORDER IN WHICH TO TEACH THE MATHS

> This chapter includes:
> 1 Networks of learning
> 2 The need for progression
> 3 A long-term maths plan for the whole school

There are some parts of the maths curriculum that seem to be obviously in a hierarchy, so we expect children to be able to work mentally with numbers up to ten before they work mentally with numbers up to 100, but many things seem not to be in a particular order. Later in this section, I will suggest some things that do seem to come in some kind of hierarchy, but I am starting this section with all the health warnings about why this is rather a dangerous idea if we overdo it.

NETWORKS OF LEARNING

Our scheme of work needs some sense of 'A comes before B', but it is also an area of danger. If we get too obsessed with this, it can be very time consuming and unnecessary because children seem not to learn in a linear way, i.e. they don't all always learn A then B then C then D, etc.

FIGURE B18 *If maths were learnt in one set order, children would be able to learn concepts like this in increasing difficulty.*

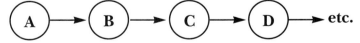

However, the reality in the classroom is that it seems much more like A, then a bit of B and a bit of E, then some G and a bit more of B, then some J and a smattering of M and a bit of C, D, E and F, then when we teach a bit of H, you discover that they didn't learn that but instead picked up quite a bit of I and J along the way (see figure B19). No wonder teaching is so complicated! (If you want to read more about networks of learning, you can read Denvir and Brown, 1986.)

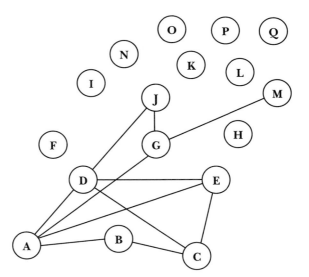

FIGURE B19 *Children's learning is better described as a network than a line.*

What seems to happen is that when we set out to teach a particular aspect of maths to a child, they don't always learn what we set out to teach, but they often learn other, related things. Children seem to learn in a complex way. They seem to build on what they already know in a way that we cannot always predict. They often learn things that we might have thought they were not 'ready' for or that are huge leaps ahead of where we thought they were.

One thing that comes through in the observations from the CAN and PrIME project (1986–90) was the impression that we can often underestimate what children are capable of. For example, it is very common to find reception children kept to numbers one to five (not even zero to five) in the first term and up to ten in the second. Although this makes sense at one level, this is restrictive, and small children enjoy working with the idea of millions or infinity. We wouldn't necessarily expect them to calculate with large numbers at first except on a calculator, but by being restrictive and putting maths into bite-sized portions, we fail to reveal the excitement of maths and the adventure of exploring interesting things, such as huge numbers or a seven-year-old asking why thirteen divided by two gave what he thought was sixty-five on the calculator.

More questions of what comes first

Some things are argued about in maths education concerning the order in which we might do things, for example do we teach two-dimensional shape before three-dimensional? The arguments are something like this.

NUMBER	OPERATIONS				FRACTION				
	addition	subtraction	multiplication	division	common	decimal	percentage	length	
• 1–1 correspon-dence • more than/less than • counting to 10/50/100 • use of symbols for numbers • ordinal numbers • calculator numbers to 100 • place value to 50 • place value to 100 • counting to 1000 • place value to 1000 • calculator numbers above 1000 • recognise and write numbers above 1000 • multiples • factors • prime numbers • ratio and proportion • negative numbers	• counting • + to 10 • mental + to 20 • + to 20 • using calculator to relate + and − • + and − to 100 using a variety of apparatus • mental + and − using a variety of methods • exploration of + and − beyond 100 on calculators • patterns in numbers • + and − of large numbers beyond 1000 • using calculator to perform complex + and −	• − to 10 • − to 20	• grouping into equal groups • use of constant on calculator • counting in 2s, 5s, 10s • language of 'sets of', 'lots of', etc. • × by 2, 5, 10 • use a variety of methods to multiply • link between × and ÷ • × by 100, 1000 • × by numbers 2–10 • relating × to ÷ • × by numbers 11–20 • × large numbers by calculator • × large numbers using a variety of methods	• early ideas of sharing • sharing into groups • division by repeated subtraction • use of ÷ on a calculator • sharing into groups with remainder • recording 'short' division • divide large numbers using a variety of methods	• idea of half/ quarter • early ideas of equivalence • use of thirds, sixths, etc. • notation of $1/2$, $1/4$ • fractions of objects, shapes, numbers • equivalence • idea of $1/3$, $1/6$, $1/9$, $1/2$, $1/4$, $1/8$, $1/16$, etc. • use of $1/2$ as 0.5, $1/4$ as 0.25 on calculator • addition/ subtraction of fractions • multiplying and dividing fractions	• link with fractions using calculator • use of money to show decimals • use of 1/10ths • + and − • use of 1/100ths • × and ÷ • recurring decimals	• use of % in real problem situation • link of % to decimals and fractions	• comparison of lengths • languages of length • using non-standard measures • standard measures (cm, m) • use of mm and km • + and − of linear measure • scale • using maps and plans • using cm as decimal fraction • perimeters, circum-ference of circle, etc.	

FIGURE B20 *This chart shows in outline an attempt to put maths in some kind of order. It is far from complete, and you will need to fill in the details in your scheme of work.*

MEASURES						SHAPE AND SPACE	
weight	money	volume/capacity	time	angle/movement	area	2D/3D shape	movement (see angle/movement)
• comparison of weight	• uses of coins in shopping	• sand/water play	• use of language of time: fast, slow, morning, seasons, etc.	• use of $1/4$ turn in PE	• early ideas of covering a surface	• sorting 2D and 3D shapes and describing	
• language of weight	• equivalence of coins	• language of comparison (full, empty, etc.)		• use of forward, backward, left, right with turtle	• comparison of areas	• using language, edge, straight, curved, etc.	
• using non-standard measures to balance	• giving change		• measuring passage of time with tockers, sand timers, etc.	• use of $1/4$, $1/2$ turn with turtle	• use of non-standard units		
• use of standard measures (g and kg)	• + and −	• using non-standard measures				• using properties of shape to build and draw	
	• using money for place value	• using l and ml	• reading hours on analogue and digital clocks	• use of $3/4$, whole turn with turtle			
• + and − of g and kg	• equivalence of coins to £1	• making nets of shapes	• reading time in hours and minutes on digital clock	• use of right angle		• naming common 2D and 3D shapes	
	• shopping over £1	• volume of shapes with cubes		• recognising turns in clocks, doors, etc.			
	• + and − with large amounts	• use of 1cc for volume of shape	• reading half past on analogue			• language of properties of shapes, including irregular shapes	
	• × and ÷			• using Logo to draw	• tessellation to cover an area		
	• decimal notation with money	• using ml as decimals of litre	• knowledge of months, seasons, days of week, year	• using Logo to draw 2D shapes	• using grids to measure and compare area	• tessellations	
• using g as decimal fraction	• complex +, −, × and ÷	• volume of regular shape	• using units of time	• use of degrees		• nets of 3D shapes	
• metric tonne	• use of percentage	• volume of irregular shape	• reading hours and minutes on analogue clock	• drawing and measuring angles with protractor	• using cm^2 to measure area	• reflective symmetry	
					• areas of regular shapes	• finding several axes of symmetry	
			• 24-hour clock	• acute obtuse, reflex angles	• use of hectares	• rotational symmetry	
			• use of minutes and seconds to measure time	• angles in regular and irregular shapes		• translations	
			• $1/10$ and $1/100$ of seconds			• congruence of shapes	
						• geometrical constructions	

- Two-dimensional comes first because these shapes make up the three-dimensional shapes.
- Two-dimensional shape names are easier to learn.
- Two-dimensional shapes can be made by the young child.
  or
- Three-dimensional shape comes first because a child lives in a three-dimensional world.
  or
- We decide to come up with some compromise and teach them together. That is my own way of teaching shape, but you will need to decide for your scheme of work how you want to do it in your school.

Some things do seem to come in an order

Although it is very difficult to put maths in some kind of order, it is possible to draw up some kind of very general sequencing of some topics, and I have given an example in figure B20. You don't have to agree with it! I'm not at all sure that we know enough to be sure about much of this e.g. children of seven deal perfectly well with negative numbers if they have a calculator.

THE NEED FOR PROGRESSION

Unless we get some sense of what we will teach and when, we can end up with a scheme of work that is weak on continuity and progression. What happens to topics like tessellation, fractions and symmetry?

It is possible to envisage some maths, such as tessellation, fractions, symmetry, measuring the playground with a trundle wheel and making boxes with Polydrons, going on in classes with the very youngest children and with the oldest children. That would be expected in some ways. It is often said that the art of being a good primary teacher is to find a hundred different ways of teaching the same concept. However, when we write our schemes of work, it is crucial to try to build in some way of differentiating some of these tasks so that the six-year-olds will have a different focus on their tessellation work than the eleven-year-olds, and this is where I find the idea of some kind of hierarchy in maths quite helpful. Let's look at some examples of this.

Tessellation

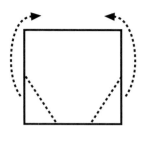

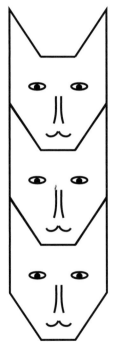

FIGURE B21 *Starting with a square, cut off the two lower corners. Then, without rotating them, stick them onto the top of the square. The shapes will tessellate. This works with any regular shape that tessellates (e.g. hexagrams and triangles).*

FIVE- AND SIX-YEAR-OLDS

The focus of this work might be:

- to find out by playing with them which shapes fit together;
- to learn the names of some of the shapes that tessellate and be able to discuss some of the properties of these shapes (these have curved edges and these have straight edges, etc);
- to make some patterns and start to appreciate some of the aesthetic qualities in maths;
- to start to get a feel for some of the properties of two-dimensional shapes;
- to make up some groups of shapes (these ones tessellate, these ones don't);
- to work with the common geometric shapes;
- to experience enjoyment with a mathematical task.

SIX- AND SEVEN-YEAR-OLDS

The focus might be:

- to try to cover a piece of paper leaving no gaps and decide which was a good shape to do that with;
- to look at shapes that fit together in nature;
- to begin to get some experiences with covering a surface as a preliminary to working on area (we needed twenty-six of these little shapes to cover this paper, but only six of these big ones. Why was that?);
- to work with sets of tessellating shapes that include shapes such as arrow head and crosses to widen the experiences of tessellation;
- to tessellate hexagons on the screen using Logo and build a procedure to do this.

SEVEN- AND EIGHT-YEAR OLDS

This might develop to be:

- continued use of Logo and increasingly complex procedures;
- a knowledge of which common geometric shapes tessellate;
- using shapes in a variety of problem-solving situations (where are there rectangles in the school that fit together? Can you make a tessellation design on your art folder? Can you make a net for a cube? How many different nets can you make for a cube?).

EIGHT- TO ELEVEN-YEAR-OLDS

This might develop to be:

- taking a shape that tessellates and changing that shape in some way so that the resulting shape will still tessellate (see figure B21 on page 61);
- discussing what properties of a shape make it a shape that tessellates;
- making Mottik tessellations;
- creating their own patchwork pattern to sew or stick;
- understanding the use of square centimetres for covering a surface as a useful way to measure area;
- explore ideas such as 'do all triangles tessellate?' 'Does any quadrilateral tessellate with itself?';
- appreciating that if you wanted to measure the area of the playground, you would probably not use the same size unit as when you measure the area of the surface of a table.

So, although children may encounter tessellations in almost every year of their schooling, at each stage it develops a bit (as well as re-capping on what has been previously learnt) and so there is a real sense of progression.

Making boxes with Polydrons

FIVE- AND SIX-YEAR-OLDS

This might involve:

- playing with Polydrons in free play to begin to appreciate some of the attributes of the shapes and to learn to manipulate the materials;
- talking about what they have made at 'review' time;
- counting the number of the shapes they have used and the names of the shapes ('I have used six triangles to make this shape');
- developing a sense of the ways in which two-dimensional shapes can make three-dimensional shapes and structures;
- completing some design task, e.g. making a garage big enough to take the tip-up truck;
- estimating how many shapes they might need for a task before they start.

SEVEN- AND EIGHT-YEAR-OLDS

This might develop to:

- knowing the names of and being able to construct some of the common three-dimensional shapes, e.g. tetrahedron, square-based pyramid and triangular prism (see figure B22);

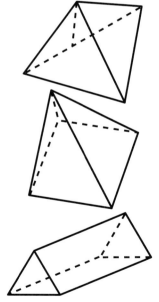

FIGURE B22 *Common three-dimensional shapes, e.g. tetrahedron, square-based pyramid and triangular prism.*

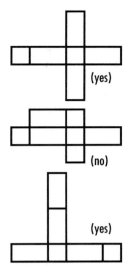

FIGURE B23 *Will these nets make a cuboid?*

FIGURE B24 *A triangular Polydron with layers around it.*

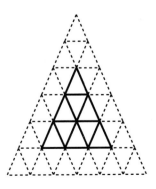

FIGURE B25 *A growing pattern with triangular Polydrons.*

- being able to make several different nets of common shapes, such as tetrahedron and cuboid, and being able to tell which ones will not successfully make up into the required shape (see figure B23);
- using the shapes to make patterns that can be explored as a part of algebraic understandings about growing patterns and developing the child's ability to generalise. In figure B24, the triangular Polydrons are arranged in layers around one central Polydron. How many triangles in the next layer? Can you see a pattern? (The beginning of generalising.) How many do you think would be in the fiftieth layer?

NINE- AND TEN-YEAR-OLDS

By this time, children would be using Polydrons for specific tasks that involve them in some kind of problem solving or investigation such as:

- how many of these square Polydrons do you think you would need to cover the carpet? You can only have five Polydrons to help you;
- can you make a growing pattern with Polydrons for your friend to work with?
- use triangular Polydrons or Mottik to explore this pattern (see figure B25).

What these examples are intended to show is that it is important to build into the scheme of work some sense of development of mathematical activity, so that teachers who feel they need support with their maths are not left wondering how what they teach in symmetry is any different with their eleven-year-olds from what they saw the six-year-olds doing last week.

That progression needs to be there explicitly, and it will be clear from the examples that I have used that most teachers find that this kind of repetition and failure to show progression in tasks seems to be harder in areas of maths that aren't about number. (But it is also true for fractions, as was noted in the OFSTED report of 1994, page 9.)

The various curricular documents are maybe much clearer about the progression in number, and most of us can grasp that the two, three and ten times tables are easier than the seven and nine times, so we know that a child of five or six learns to count in twos, threes and tens long before they learn to count in sevens.

A LONG-TERM MATHS PLAN FOR THE WHOLE SCHOOL

At some stage in thinking through your scheme of work you will want to plan, to some extent, what the children do in each year so that there is less needless repetition and to ensure that each child covers the whole curriculum and gets enough repetition to be revisiting areas regularly, without the boring kind of repetition that switches children off and really gets parents down.

Schools have a variety of ways of planning this, and these are often put up on the wall in the staff room for easy reference, along with the overall plans for other areas of the curriculum. A maths plan for the whole school might look something like figure B26. This hasn't got all the detail added by teachers over the two years that it had been in use, but it shows how each teacher can know which maths topic each class is doing. They shared around the maths equipment, so that if you wanted something like the accurate stop watch unexpectedly, you knew who had it and could ask to borrow it. Each teacher also did number work throughout the year in some form or other, and they used numerous games to reinforce concepts and skills learnt previously.

> - Photocopy sections of this chapter for staff to read as the basis for discussion.
> - Present some of the ideas on progression to the staff and get small groups to look at some other areas of the curriculum in this way.
> - Analyse what is done in small groups and prepare an action list for how you could incorporate your ideas on progression into your scheme of work.
> - Decide on some principles rather than a rigid format.

	Autumn term	Spring term	Summer term
year 3	• place value • multiplication (2×, 3×, 5×, 10×) • 2D and 3D shape • symmetry	• measuring/ data handling • money and time, use of stopwatch	• maps and co-ordinates • weather and use of datebase/ volume and capacity
year 4	• fractions • place value, multiplication and division • 2D and 3D shape, tessellation, printing symmetrical designs	• area and volume • Logo, angle, movement	• multiplication, division • designing patterns using parallel lines • use of compasses for circle patterns • temperature • weighing • use of all standard metric measures
year 5	• fractions and decimals • use of calculator for decimals • place value to a million	• area – tangrams	• decimals
year 6	• Logo, data handling • fractions and decimals • tessellations of complex shapes • construction of 3D shapes	• graphs • volume of 3D shapes • capacity, weighing, money, length	• revision of all multiplication tables • revision of all four rules

FIGURE B26 *It is helpful to make a long-term plan for the whole school, so that you can be sure of continuity and coverage, and also to share equipment. You can put a plan like this on the staffroom wall and leave space for additions and changes.*

CHAPTER 6

HOW LONG TO SPEND ON THE MATHS

> This chapter includes:
> 1   Case studies and commentaries

CASE STUDIES AND COMMENTARIES

INSET In this section, there are a number of very brief case studies of the ways in which various schools plan their maths and the amount of maths that they have agreed each child should do. You could use this section as a basis for discussion during INSET.

The Dearing Final Report of December 1993 (published 1994) recommended:

KEY STAGE 1
- Time spent on maths to be between sixteen and seventeen per cent.
- A total of 126 hours a year.
- That is about three and a half hours a week.

KEY STAGE 2
- Time spent on maths to be between fourteen and fifteen per cent.
- A total of 126 hours a year.
- That is about three and a half hours a week.

School 1

A country school with a varied catchment area.

MATHS DONE

For one hour every day, each child worked from the scheme on their own. The children planned when they did this during the day. Tasks were put on the board (usually about three or four a day). The children did these in any order.

COMMENT

- A lot of time on maths, maybe not that balanced.
- What about the role of talking in maths?
- Gives the impression that maths is a silent task that you do on your own.
- Could be relatively 'teacher free', but in practice often quite teacher intensive if you want to do it well.

School 2

A large city first school on a large housing estate.

MATHS DONE

Each class had a suggested format of the day and teachers felt that they had to stick to this format.

9 – 9.30	Shared reading with parents.
9.30 – 10.40	Register (so they could do five minutes' mental maths) and then group work that could include maths along with writing, science, technology, etc.
11.00 – 11.50	Every child had to 'get on' while the teacher had one reading group with her. In reality with early years children, this often meant that they had to be 'choosing', and although this could be well structured on the 'plan, do, review' framework, with each child choosing an area of work and reviewing with the rest of the class in the last ten or fifteen minutes, it was not a time that a teacher could be closely involved with any maths work. Maths had to be completely 'teacher free'. (Where children initiated this work or where they knew maths games quite well, or were working with their calculator, this worked well. However in Key Stage 2, where maths was done entirely individually from the scheme, there were many problems.)
11.50 – 12.00	Review, story, clear up.
1.00 – 1.30	Every child did silent reading.
1.30 – 2.30	Group work time (so this could include some maths).
2.30	Clear up and review time and then story.

COMMENT

Maths had quite a low profile compared with reading (the school has children from three to nine years), and when the maths co-ordinator tried to build up a maths games library so that children took home some maths as well as their reading book, the head was obstructive, and although teachers and some parents wanted that and worked hard to make the games, the idea never got off the ground. Teachers thought that each child did between two and four hours maths a week, but when three teachers agreed to time how much maths some individual children did, they actually found that it varied from just an hour for some children and the maximum that any child did was one hour forty minutes – in a week! That's not nearly enough.

Fitting maths into this system means that considerable work with the whole class needs to be done (e.g. showing everyone how to play a maths game or giving a verbal open-ended starting point). Several of the teachers felt that they did too little maths, but within the format of the day and with all children working an integrated day, it was difficult to give maths a higher profile.

If you are in a setting like this where there is too little emphasis on maths, it can be difficult to raise its profile. See page 136 for suggestions.

School 3

A large, three-form entry school on an inner city housing estate with over fifty per cent of children who do not speak English as their first language.

MATHS DONE

In this school, teachers worked from a scheme, but they moved around the scheme according to the maths topic being taught. They did not go from page 1 to page 2, etc. through the books, but did their maths topic for the half term (such as fractions) by using the scheme and other resources.

- Children always worked in ability groups.
- Maths was on Monday, Wednesday and Friday, so three hours a week.
- Not much arose from 'topic' teaching, but there was some mental maths, so each child maybe did about three and a half hours a week.
- No home/school maths.

COMMENT

Heavily scheme based, but this worked well, as the scheme was a good one that provided many open-ended starting points for the whole class, with children doing different sections of the books depending on how well they responded to that initial task. Some of the staff used other resources to supplement the scheme and they liked the way that the scheme taught in maths topics such as two-dimensional shape. They felt that the scheme gave the children a balanced diet, especially as it had lots of games, investigations and also starting points for computer and calculator work.

School 4

A suburban school with very varied intake.

MATHS DONE

This school had a maths session each day, usually between play and lunch. Three year classes were split into four ability groups, with each teacher staying with their group and the head taking one group.

Each child did about five hours a week. There was a home/school project, and teachers also linked in maths where they could in their 'topic', e.g. one class was doing Victorians and they designed a time line around the class where one year was 1 cm. The children planned this themselves and went on to do other things to scale, e.g. the plan of a nearby village.

COMMENT

The staff did feel that the ability groups were a bit rigid, but teaching in these groups made planning easier because there was a smaller range of ability to plan for.

Maths had a high profile, and the head taking a group was strongly appreciated.

You can see from these examples that time spent on maths varies enormously in different schools.

- Get teachers to outline their practices of how much time they spend on maths. Critically review this with the whole staff.
- You could use the table in figure B27 as a starting point for discussion.

JUNK FOOD	JUNK MATHS
There is a lot of it about.	See most school textbooks.
All the preparation is done for you.	This is done for you by the authors or the teacher.
The instructions are simple and laid out in steps so that you don't need to think or be creative.	See most textbook questions.
It is superficially attractive, but turns out to lack flavour.	It looks well structured and appears logical, but it can be dull and lack substance.
It does you little good. It tends to pass through you quickly.	Pupils can become unable to retain their maths or apply it in new contexts.
All the real nutrients are removed and substitutes have to be added.	It offers few real-life situations, but invents and contrives them.

Danger: health warning

Junk maths can seriously damage your pupils.

FIGURE B27 *I have adapted this from a chart in* Better Mathematics, *HMI, 1987. The bit about junk food passing through people quickly doesn't seem right – I thought it went through rather slowly – but the link with maths is, I think, helpful and can provide the starting point for discussion about the kind of maths 'diet' that we want our children to have.*

SECTION

PLANNING FROM THE SCHEME OF WORK

This section is divided into three chapters:

CHAPTER 7
Medium-term planning

CHAPTER 8
The use of resources

CHAPTER 9
Assessment and record keeping

MEDIUM-TERM PLANNING

This chapter includes:
1 Starting with maths topics
2 Starting from a cross-curricular topic
3 Combining both approaches
4 Ongoing maths
5 How much of each kind of maths?

Four models of planning maths for an individual teacher will be offered, showing how these might relate to the whole school planning.

STARTING WITH MATHS TOPICS

(For example, place value, three-dimensional shapes, or other topics from Programmes of Study in the National Curriculum.)

School A

An inner city school with over forty per cent of children from non-European cultures.

This school used to use a commercially produced maths scheme, so that each child was used to working each day, on their own, at their own pace. Maths was seen by the children as something of a competition to get ahead of other people, and the parents, too, saw maths as the scheme.

A new teacher at the school could see that the standard of maths was very low and that children did not really seem to enjoy the subject. She changed how the maths was organised in the class and worked with very loose ability groups for 'exposition' times; for example, when doing place value as a topic, different ability groups discussed different levels of tasks together, but all the children could choose which pages and which books they did their consolidation from. The scheme books needed to be retained because of the school policy and because the children

	RED	YELLOW	GREEN
Investigations	Choose from:	1) The desert problem 2) Cutting out half 3) Magic squares 4) You can choose from the red box	Finish your spirals. What if you did the spirals on hexagonal grid paper?
Computer	Logo groups – see timetable. Finish putting your EDITOR information on database.		
Design & technology		1) Design a bird table for a family that has cats. 2) Finish your working models. 3) Finish your book cover. 4) Finish symmetrical prints.	
Tables	2x 5x 10x	revise: 2x 5x 10x new one: 3x, do on a 100 square	3x 6x 9x Do digital roots investigation.
Games	• Rocket base 10 • Play any place value game backwards • Build a cube bases 3 or 4 or 5 • Finish your own game • Choose a calculator game from the yellow box	• Build a cube and shrink a cube • Race across the board • Finish your own game → Make your own subtraction game	• Finish your place value game → Make a place value game with a calculator
Scheme	Book 1 pp 16, 17, 18, 19	Book 2 pp 16, 17, 20–24 Book 3 pp 14, 15 Use the Dienes!	Book 3 pp 14–21 Book 4 pp 14–19 Try anything you want from books 5 and 6

FIGURE C1 *Scheme of work, format 4.*

saw this as 'real maths' (as did the parents), but the teacher worked to plan a more varied diet for the children and not just maths from a scheme.

Work for the year was planned around maths topics, e.g. shape and space, subtraction, movement, angles and directions, based on the Programmes of Study (see Figure C1). In reality, it was found best to mix these topics a bit, so from October to December place value was the topic (incorporating subtraction and addition), but each fortnight a different aspect of shape was included in the planning, as this allowed the children to do some art work (something aesthetically pleasing), and technology (using the computer well).

The groups were often the groups that the children were taught in, but they were very flexible indeed, enabling the children to attempt the most challenging work if they wanted to – and several did, successfully. Much of the work was introduced to the whole class (games, investigations, designing, etc.). The scheme book that they had been 'on' seemed to have little relationship with their actual mathematical ability, and by removing the scheme book as something that they did every day, the teacher thought that they began to believe in their abilities more, and maths in the class became a favourite activity.

Examples of the shape work that ran alongside the place value work in the autumn term were:

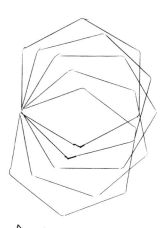

Abdul   (age 9)

FIGURE C2 *Rotating a shape.*

- making circle patterns (this gripped the children so much that they produced these at home for the next few months. Work was put on display and put into a class floor book – a large book kept in the reading corner that children could contribute to when they wanted);
- patterns with parallel lines;
- making cut-up tessellating shapes from squares, rectangles, etc. (see the tessellating fox on page 61);
- what happens when you rotate a shape (see figure C2)?
- as Christmas approached, working on three-dimensional shapes and constructing them from scrap card for tree decorations;
- designing a star including three-dimensional stars;
- ongoing use of Logo on the computer, with each child working in groups of two or three throughout the day. The computer went on at 8.30 a.m., and groups were planned throughout the day, often including assembly times and PE, where that could be arranged, so that every child had a substantial amount of Logo time every two weeks. There was also computer club at lunch time and sometimes after school;

- ongoing work using a home/school project that included shape ideas.

HOW MUCH OF EACH KIND OF MATHS?
(See also page 80.)

In terms of the amount of time that each child spent on each aspect of maths over the fortnight, this varied enormously depending on the tasks that the child was doing, but each fortnight the teacher started each group off with some kind of open-ended problems, such as subtractions squares (see page 52). So every week the children did some kind of problem solving or investigation. Games were a central feature of the teaching, so most children would play some kind of game each week. Maths was a central feature of 'review' or carpet times, when groups or individuals shared what they had done in maths.

Most children did some work from the scheme in a fortnight, but not always, especially as the term progressed and children became so involved with the investigations that their wish to keep doing the pages from the scheme diminished. Mental maths was done each day, and the teacher also used the blackboard for some 'quickies' that the children sometimes did when they first came into the classroom in the morning. (This was mainly a device for settling the children down and was used alongside other tasks for the start of the day.)

School B

Another school based their scheme of work on the five areas of maths outlined on page 29 and taught from maths 'topics'. Their topics included all the things that you would expect listed in the content list plus the following:

- maps and co-ordinates (they did this through pirate maps in their topic on 'explorers');
- orienteering (they went out into the surrounding area for this);
- maths from a spreadsheet for older children;
- they taught much of their shape work from Logo (they had a computer in each class and every child had at least three hours of Logo in a block each half term);
- puzzles (there were interactive display up with puzzles and challenges that were changed each fortnight);
- games (every class had numerous maths games and some of these were rotated around the classes each fortnight. The games covered most of the maths curriculum. Each class had a set of place value games that they sometimes swapped with

other classes but which they expected the children to play throughout the year in the belief that place value is so crucial that the children needed to have constant reinforcement of the concepts).

The games were listed into the scheme of work so that they were used to both teach concepts and to reinforce them. Figure C3 shows a simplified outline from one of their pages about number.

LEVEL 1	MATHS TOPIC: Number		
Programme of Study	Activity	Resources	Grouping
• develop flexible method of working with numbers orally and mentally	• spotty monster game	CUP games box module 1	pairs
	• mental maths	Teacher's book, pages xx and xx	whole class
	• calculator activity xx	BEAM calculator pack	small groups
	• 'Bounce to it', page 7, 'Arithmogens'	'Bounce to it', cubes, calculators	whole class, then small groups
	• Homelinks activity number game	Homelinks books	whole class introduction

FIGURE C3 *Scheme of work, format 5. This has been simplified for reproduction here.*

ADVANTAGES OF THIS WAY OF PLANNING
- What is clear about this way of planning maths is that some elements are crucial. We must plan to cover the whole curriculum.
- It is easy to plan as you use the curriculum as a starting point.
- It allows a systematic approach to planning so you can be sure that you have covered all the content.

DISADVANTAGES
- It could get a bit 'dry' and narrow.
- It could lead to fragmentation and compartmentalising children's experiences.
- It might not make the best of cross-curricular links that could be one way that a child might see the relevance and meaning in maths.

STARTING FROM A CROSS-CURRICULAR TOPIC

(For example, ourselves, or Victorians, or starting with a problem of some kind of very open task that might incorporate a range of the elements of many Programmes of Study.)

If you were doing Victorians with nine- and ten-year-olds, you might do:

- something on 'old' money and imperial measurements involving different bases (e.g. base twelve with twelve pennies in a pound);
- some census work from a computer data base;
- a map of the Victorian parts of your town;
- tables by rote and have lots of fun chanting them.

You can start with some kind of 'open' task, such as planning a picnic, or asking children to 'explode a number' (see page 142) or to find the volume of air in the room or suggest that, if taking the guinea pigs home is causing a problem, they could plan a timetable so that they take it in turns to make it fair.

ADVANTAGES

- It is important to get maths from a topic when we can because it shows maths in a meaningful context and it can show the use of maths in everyday life; for example, building a Viking ship to scale involves essential measuring skills, cooking a medieval banquet involves planning, etc. Science, geography and technology, etc. include considerable mathematical ideas that we can draw out. For example, a graph of how far the truck went when we had the slope at different angles is a perfect data handling task. For both science and maths, it is crucial for the child, from the very beginning, to start to interpret their graph and raise new questions. Drawing a pretty graph and putting it on the wall is not the end of the task because graphs are about communicating something. So what has it communicated? What can it tell us? Are the results what we expected? This kind of maths can have enormous meaning for children.
- Planning from this kind of more open task can mean that you will be addressing many different parts of the Programmes of Study in one activity.
- Starting from an 'open' task, we can give one task to every child in the class, and the differentiation will be in the outcome, not through the tasks we plan for each ability group. This can help enormously with classroom organisation and

allows children to achieve things that we might not expect from them. (When the children mentioned on page 15 planned their sports day, the teacher assessed the outcomes of their work. They all had the same task, but the outcomes were hugely varied. When a teacher of five- and six-year-olds did a topic on 'ourselves', some children were able to put information on each child into the data base independently (heights of children, length of hair, shoe size, the number of brothers and sisters, etc.), whereas some of the children needed considerable help to do the measuring in the first place and to enter it on the computer. All the children were able to interpret the graphs that resulted from the work, but some of the more able ones asked if they could now measure some of the children in the nursery and reception class.)

DISADVANTAGES
- The problem is that if we try to get all the maths from the topic, we might miss out some bits of the curriculum, and children may not pick up a coherent understanding of the whole curriculum because, inevitably, when working from class topic areas, the maths will be more *ad hoc* than if taught from a mathematical topic area (such as data handling or multiplication).
- You could find some difficulties in keeping track of exactly which child has done what. (But given the way that children seem to learn in networks, we need to be realistic about the amount we can actually keep tabs on who learnt what, even with a very tight system of planning and record keeping.)

COMBINING BOTH APPROACHES

Finding some kind of middle position that uses the advantages of starting with the Programmes of Study and also the advantages of starting with good open activities and those that link to the wider class topics, would seem to be the ideal. Probably most of us will use some combination of these three methods of planning. So we might plan something like 'water' as a topic and then teach, within that, volume and capacity.

You could start with something like figure C4 as an outline plan for some of the topics you might want to teach. You could search various resources to find some good activities, and all the children in one year group might work on 'water' but experience a variety of activities that also cover all the aspects of measuring volume and capacity suitable for their level of ability.

- Comparing and ordering containers using appropriate language.

- Using non-standard measures, e.g. cupful.

- Conservation tasks, e.g. tall/thin/wide containers.

- Estimations of capacity.

- Using standard measures (including imperial).

- Sinking/floating activities (see science document).

- Making boats and testing how much load they will carry (see xxxxx book, page 16).

- Measuring depths using the depth gauge.

- How much water is in the water bath?

- How could you measure how much liquid you and your family drink in a day?

Water/Volume/Capacity Key Stage 1

- Displacement activities (use displacement bucket).

- Finding volumes of Polydron 3D shapes. (Could use buttons, pasta, etc.)

- Does a yoghurt pot of water weigh the same as a yoghurt pot of sand?

FIGURE C4 *Topic outline plan*

Of course, this has disadvantages (you need to check carefully that all the curriculum is covered), but it might be possible to have all the advantages of the two extreme positions above.

ONGOING MATHS

(For example, mental maths, number, measuring and data handling.)

Integrated within whichever one of those three possible ways of planning we choose for our scheme of work we might decide to have some kind of basic ongoing maths in the classroom all the time. (So if you are doing a topic on water that focuses on the maths of measuring capacity, you might want to keep number going on in some way, plus making all the links to shapes that you can.)

ADVANTAGES

Keeps number ideas (or some other area of maths) ticking over while there is a concentration on teaching either another area of maths in detail, or during the days and weeks when you are having a science or language 'blitz'.

DISADVANTAGES

These depend on how you do this basic maths. If you are doing maths just from a scheme, there is the disadvantage that it is boring, often far too easy for the children, as outlined on page 108.

The types of basic ongoing maths that you might want to try out are:

- Logo and floor robot groups going much of the time;
- other 'good' computer programs (open-ended), such as data handling packages, art packages or programs like Slimwam (see resources list at the back of the book);
- data handling from things arising in class (how many people going on the theatre trip?);
- mental maths covering number, and any other areas of maths that can appropriately be done mentally;
- 'register' maths, e.g. there are thirty-two people in our class and five are away, so how many are here? How did you work that out Nasreen? Did anyone do it a different way? How many different ways are there in our class?;
- teaching children a new maths game every day for a week at 'carpet time' and having a box of maths games in the class that children can choose freely from (or you can select according to your maths topic);
- children taking maths games or activities home at the weekend;
- construction activities;
- a different 'shop' each half term;
- a free-choice box of open-ended starting points that children can choose to do in their own time.

HOW MUCH OF EACH KIND OF MATHS?

The teacher in school A on page 72 had worked on this issue with her colleagues. They thought that they came to a reasonable balance of each fortnight block having:

- some consolidation from the scheme;

- about two or three investigations (but they found that these varied enormously in the time they took and the ways that they were extended, so they made the policy decision that it was better for a child to go into depth with one investigation than to do a few superficially);
- several games (often, but not only, used to occupy children usefully while the teacher was busy);
- teacher-led 'exposition' of new topics, games, or class review discussion of what had been learnt;
- a core of maths activities that varied with the topic;
- computer work built in.

With mental maths every day, this is a reasonable 'balanced diet'. Of the three to four hours of maths a week, each child did about two hours of open-ended investigations and problems, about an hour of consolidation or textbook work and about an hour of other work (computer, mental maths, etc.).

---

1 Over the space of two weeks, keep a 'diary' of the maths that your class does and take that to a staff meeting. You could discuss:
   - how balanced is my maths teaching?
   - how much of each kind of maths do I do?
   - which of the four approaches in this chapter (starting from maths topics, etc.) do I prefer?

2 You might well decide that you like each of the four approaches at different times and for different maths topics, and you might want to incorporate that into your scheme of work. To help with this, you could break into small groups and:
   - think of one activity for each of the four approaches and present them to the rest of the staff;
   - put all the activities generated into a resource bank and ask all staff to try out all the relevant activities with their class;
   - fix a date for discussing results.

Deciding to return with a list of pros and cons for each approach might be helpful. What balance of each kind of maths do we want in our scheme of work?

CHAPTER 8 THE USE OF RESOURCES

This chapter includes:
1 Apparatus
2 Setting up a resource bank
3 Setting up a maths games library and home/school maths
4 Calculator and computers
5 Published maths schemes
6 Human resources

APPARATUS

The effectiveness of mathematical experiences in the primary school depend for their success on practical apparatus. Without a good range of it, much of what we could do to improve the learning of the children is made difficult – maybe even impossible – and certainly very frustrating.

What every class should have

We should expect each classroom to have at least the following minimal apparatus. All age groups need this apparatus, so that there are always things available for children to develop their own ideas and so that open-ended tasks can take place all the time (e.g. what is the tallest thing in the school hall?). This kind of investigative maths will be severely constrained if children cannot have apparatus readily available.

- Infant classes need lots of counting equipment, such as buttons, shells, coins, etc.
- Multilink and/or Centicube, and tracks for these.
- Unifix (two trays) and associated apparatus, e.g. abacus, number lines with markers, etc.
- Polydrons, Clixi, Mottik or Jovo.
- Some kind of 100 squares (e.g. Unifix 100 square trays and 'window markers' with grids 1–100 and 0–99).

- Cuisenaire.
- Place value equipment – various bases plus base ten units, tens, hundreds, thousands (Dienes or similar).
- Place value mats.
- Number lines (0–100) marked and unmarked, portable (wipe-clean ones are ideal) and/or stuck onto every table, and a long one on the class wall.
- 0–100 cards and other packs of cards and blank cards.
- 100 squares, numbered and blank.
- Various dice, including blank dice.
- Dominoes and other maths games.
- Four-function calculators (at least ten).
- Fake money (as realistic as possible and/or some real money).
- Various measuring devices: tape measure, surveyor's tape, metre rulers, bathroom scales, balance scales, kitchen scales (press down), one of each capacity measure in metric and imperial, set of metric and imperial weights, stop watch, an analogue clock.
- A range of grid paper (squared, isometric, hexagonal).
- A range of different sized and graded graph paper.
- Access to as many maths games as possible.

What every school should have

The following is a complete list of apparatus that each school should have, arranged by maths topics. You could consider building up your resources in just one of these areas first, or you could store them in coloured boxes according to these categories. Some items will be needed more in early-years rooms, and others with other children; some specialist equipment, e.g. a thermometer, can be centrally stored.

SETS AND NUMBER
- Objects for sorting and counting (e.g. 'real' things, such as pebbles and shells, plus things like plastic sorting objects, e.g. teddy bears, small, three-dimensional animals, counters).
- Multilink.
- Unifix and associated apparatus (e.g. number lines); Centicubes and associated number lines (KS 2).
- Write-on number lines.
- Games – as wide a range as possible.
- Hoops for sorting.
- Beads for threading (KS 1).
- Pegboards and pegs.
- Thermometer.
- Fraction pieces.
- Fake money (as realistic as possible and/or some real money).

SHAPE AND SPACE

- Boxes and junk material.
- Polydron, Clixi, Mottik (Jovo for older children).
- Sets of shape pieces in two-dimensions (and/or some kind of pattern blocks).
- Sets of three-dimensional shapes.
- Straws and pipe cleaners.
- Geoboards.

- Logiblocs.
- Poleidoblocs.
- Roamer or Pip.
- 360° protractors.
- Safety compasses.
- Clinometer.
- Plumb line.
- Compass (NSEW).
- Mirrors.
- Tracing paper.

HANDLING DATA

- Dice.
- Spinners.

- Sampling bottles with beads.

LENGTH

- Rulers or various lengths.
- Tape measures or various lengths.
- Trundle wheel.

- Height measurer.
- Callipers.
- Micrometer.
- Depth gauge.

WEIGHT

- Balances with metric weights.
- Springs, elastic.
- Bucket balances.
- Kitchen-type scales (press down).

- Bathroom scales (metric and imperial).
- Spring balances.

CAPACITY

- Variety of containers, bottles, jugs, cups.
- Funnels.
- 1 litre containers.
- Containers of other sizes, e.g. half litre, in metric and imperial.

- Rice or sand.
- 5 ml spoon.
- Displacement bucket.

TIME

- Working classroom clock, either analogue or digital.
- Working clock or watch of other type.
- Old, dead clocks.

- Geared clock mechanism.
- Stopwatch.
- Water clocks (e.g. containers with holes in).
- Tockers.

AREA

- Transparent grid of 1 cm squares.
- Tangram puzzles, etc.

OTHER MEASURING EQUIPMENT

- Thermometer.
- Compass.

HARDWARE

- Computer.
- Concept keyboard in early-years rooms.
- Floor robot (Pip or Roamer).

SOFTWARE

- Counter.
- Slimwam.
- Monty.
- Logo and a floor turtle.
- Eureka.
- Junior Pinpoint.
- Database.
- Spreadsheet.
- An art package.

PAPER

- Plain white paper.
- Coloured sugar paper.
- Squared paper – 5 cm, 2 cm, 1 cm.
- Dotty paper – squares, triangles, etc.
- Large isometric paper.
- A range of different sized and graded graph paper.
- Coloured sticky paper.

GAMES

- As wide a range as possible.

SETTING UP A RESOURCE BANK

For a whole variety of reasons, some maths equipment and construction kits needs to be centrally stored so that everyone can use them.

Things that can go wrong

Some colleagues may not like the idea of sharing maths equipment that they regard as 'theirs'. Bear in mind that:

- some equipment is too expensive and specialised to keep in every class;

- having some specialised equipment (such as Quadro or a floor robot) for a few weeks helps a teacher to plan special activities for groups just for those few weeks and therefore get the maximum use of the equipment;
- children need to experience a wide variety of equipment, as this may develop their mental images (so they need a variety of number lines, hundred squares, Dienes, Cuisenaire, hundreds, tens and units boards, a selection of abaci and calculators when they are working on number tasks, and they need a great variety of spatial equipment to develop their manipulative skills and spatial understanding, etc.). It is very difficult to say exactly how children's images develop, and it seems sensible to present children with a wide variety of things to model their maths on, in the hope that at least one of the ways can be accepted into their way of viewing maths and help them to build on what they already know.

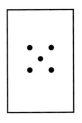

FIGURE C5 *'I always think of numbers how they are set out on dominoes, so five is always like this.' (Emma, a maths co-ordinator).*

When I was small, about seven or eight I think, I always saw numbers on this kind of track. If I had to do four add fifteen, I would see four beside me and then the track would move fifteen spaces moving behind me, so that I could see where I ended up. I still think of that now when I add. (Thomas, 19)

Ten is two of those fives, and when I add up in my head, I see numbers like that. I suppose I must have played dominoes a lot or used dice a lot. We played games like ludo and snakes and ladders a lot at home before I went to school, and I went at four, so I suppose those pictures could have been there before I went to school. (Emma, a maths co-ordinator)

School A

A new school with very limited storage space and so little maths equipment that sensible sharing was essential.

'We have our resources in each class for everyday use, and all the rest is stored in trays in the maths co-ordinator's room. It's right by the door so the children know they can creep in and get it if she is doing a story. Everything is in labelled trays, or if it is large, in boxes. Children go in pairs to fetch it, they sign it out (helped by children in the maths co-ordinator's class if the visitors are young) and then it is returned either at the end of the day or the week or half term. The maths co-ordinator knows where everything is. Her children sort it and keep it tidy and they chase up missing bits on a Friday afternoon.

It works really well. If something isn't there, you can see from the book where it is and go and ask for it. We try to plan our maths topics so that not everyone needs things like the till and play money at the same time.'

School B

A large first school that had been building up maths resources for some years.

'We have maths boxes of expensive equipment like the Polydrons, Quadro and other expensive things, like the Pip floor robot and the technical Lego and the Mottik. These boxes go into a class for two weeks and then they are passed on. The deputy head makes out the timetable, and you can ask for particular things at particular times. The children love it, as there is always something exciting arriving.

When I have these things, I use them really intensively, so I give a pair of children the Quadro or Mottik for the whole morning. I fold up some of the painting easels and put one of the tables in the corridor so that we have more space. This is particularly important when we have the floor robot, as I need to see what groups are doing and what they are learning.'

School C

A school with a new maths co-ordinator and no shared maths resources.

'We got everything out of everyone's maths cupboard on our first maths INSET staff meeting. We put it all on tables in the art area and everyone was very excited about all the stuff that we had. There was lots of 'whatever do you do with that?' and 'Oh! remember using that?' and 'Can I borrow that tomorrow?'

It showed us that it was useless for everyone to have stuff at the back of their cupboards with no one else knowing what they have, so we agreed that we would list everything and that each class needed a basic list of things and the rest would be centrally stored. We have done that now, and the problem is that some colleagues just don't use the maths store. They didn't like giving their equipment up in the first place. There was lots of 'that's mine and I don't want to share it'. Fair enough, but we need to think how to get the best use of our resources. I'm bothered that some people don't use the whole school resources, though. Some children are getting no use of things like the big construction, thermometers and stop watches. It is sad to see that expensive stuff just sitting on the shelf.

After I did the list, I wrote to the Local Education Authority and asked for a grant for more equipment, as we had decided that each class needed bucket balances and stuff like that. We were given £400 from a special fund! Well worth the time it took to find out how to get some money! The parents said that they would give us a grant, too, after the Christmas disco, so just from one evening of sorting it all out, we ended up doing rather well.'

A resource bank of printed resources

Whether you have a scheme that you like or not, you will inevitably want to use some other printed resources for maths. It just isn't possible for one scheme to give you all you need for all your children in your particular setting and your aims for their learning that term.

You can use these resources in a variety of ways, but to be useful to everyone, they need to be sorted in some way so that colleagues can see what is available; ideally, this sorting can link to the headings and format of your scheme of work. So you could cut up the resource books into individual pages according to the content and sort them into folders or boxes:

- in categories of the content of maths (e.g. place value, fractions, measuring, etc. and then maybe sub-divide this into levels of difficulty);
- in the different 'strands' of the curriculum, so you might split number into the four strands of number in Key stage 1 in the Dearing curriculum (i.e. place value, relationships between numbers (including pattern) and computation – solving numerical problems, sorting, handling and classifying data);
- in wider topics or themes, e.g. 'our school', 'about me', 'at the fairground' or 'time and seasons'.

Then individual teachers can select from the resources for their forward planning, photocopying what they need. So, using the planning sheet on page 45, they could select for a unit of work, for a fortnight or half term some games, some pages of a scheme, something from 'Bounce to it' and so on.

These planning sheets can them be put on sheets that the children can use, and you can have more than one sheet operating at any one time – maybe one for ongoing number and another one for topic-based maths. You could have different sheets for each group, or even each child when individualised programmes of work are needed. You would end up with sheets that the children keep that could look something like the one in figure C6.

Maths topic/ Programme of Study	Simple fractions: halves, quarters			Name: Sally	
Starting points	**Date started**	**Date completed**	🙁 😐 🙂	**Comment**	
• sharing out 16 play dough cakes between 5 people	15/15	16/5	😐	Sally used the words 'a quarter' to describe what she did.	
• Scheme book pages 14 and 15	17/5	23/5	🙂	Sally did this with ease.	
• BEAM geoboard activity	27/5	27/5	🙂	Clear understanding of halves and quarters.	
• 'Party cakes' game	27/5	27/5	🙂	Able to say that two quarters are the same as a half.	

FIGURE C6 *Scheme of work, format 6.*
A flexible style of sheet that the child can keep in their tray.

The starting points vary according to the choice of the teacher and their preferred teaching style. Some activities might take some time to complete, e.g. an investigation, and other things such as games and computer activities can be done more than once if that will help to consolidate learning. The child can make some kind of self assessment in the 'smiley face' column, and the teacher (and parent or other classroom helper) could make a comment, observation or assessment in the last column.

These sheets can be kept by the children while that unit of work is in progress, leaving the teacher free to observe, assess, diagnose any problems, think when to intervene and decide whether an activity still needs to be done or whether a new activity needs to be slotted in (and of course, incorporating the child's own ideas) and so on.

ADVANTAGES
- It frees the teacher.
- The child is involved in his or her own assessment.
- It provides an ongoing record of achievement and assessment.
- It is easy to operate once the initial planning is done.
- It can make the best use of the best resources and can give a child a 'balanced diet' of maths.
- It is very flexible and allows for teacher and child creativity.

DISADVANTAGES
- It needs a good resource bank of materials to start with. (Suggestions for this are in the resource section at the back.)
- The planning stage for the teacher could be time consuming, although this would get less as familiarity with the resources grew.
- The initial organisation of the resources could take some while and it would need to be done *after* there was some agreement on the format of the school scheme of work.

CASE STUDY

Just as with the scheme of work that the teacher used on page 73, this kind of scheme could well be developed a bit at a time as it is used. So if you are doing counting to 100 or 1000, you would develop this for everyone else as well as for yourself. A colleague working on two- and three-dimensional shape this term would develop that for themselves and everyone else.

Of course, this is slow, and you might want to do it faster than that. This is what one school did over a term. The head was unhappy about the ways in which the staff focused most maths on the scheme, and she thought that the children did too little practical work and far too much recording. Most of the staff agreed with this, but they were all so busy that it was by far the easiest thing to do. They decided that they needed to write a new scheme of work, but they only had a term of staff meetings in which to do it. They first decided on the areas of maths that they thought the National Curriculum covered. They decided on:

- number;
- pattern/algebra;
- measures;
- information handling;
- space and shape;
- investigation and problem solving.

The staff then divided into three small groups and took two of the areas each to work on, so one group did number and pattern, another did measures and information handling and the last did space and shape and problem solving. Then each group had a piece of A1 paper for each of their mathematical areas and for each Key Stage (see figure C7).

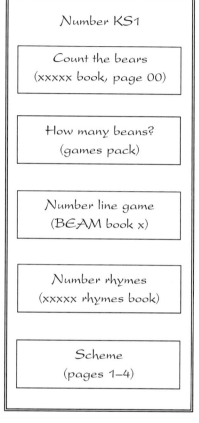

Number KS1

Count the bears
(xxxxx book, page 00)

How many beans?
(games pack)

Number line game
(BEAM book x)

Number rhymes
(xxxxx rhymes book)

Scheme
(pages 1–4)

Number KS2

Times table game x
(games pack 4)

Thousand game book 3
(pages 16 & 17)

FIGURE C7 *Making a scheme of work from a varied resource bank of printed materials. You need to give a clear reference to specific books with each activity.*

They then went through the varied resources in the school, including their basic scheme, and wrote down activities on slips of paper and blu-tacked them onto the large paper, giving the title of the book or pack and the page number.

Then, when they were fairly sure that they were getting some kind of reasonable sequence that the activities could go in, they stuck them down on the paper with glue sticks. (But they realised that they had to be careful not to be too prescriptive about this and they included huge numbers of open-ended activities so that children could explore maths, not just do tasks.)

Of course, they kept finding that they had activities to contribute to other groups, so when they found these things, they put the activity on a slip of paper and gave the paper and the resource book to that group to decide where it would or could go. At the end of each hour of work, they all stopped and looked at what each group had done and talked through their work to see if everyone felt they were still on the right lines.

The papers were put up on the staff room wall during the week so that activities could easily be added and so that everyone could look in detail at the lists. The head found a small amount of money for some new books and activities that some of the staff had seen at the maths centre, and a young member of staff had been on a course and had ordered some personal maths resources for herself which she shared with the others.

Each group had their copy of the curriculum to guide them to make sure that they were covering all the Programmes of Study, and the school record-keeping system that they had developed when the curriculum had first been published was very detailed, so they were able to check that they covered everything.

After three hour-long staff meetings, the papers were ready for typing. They abbreviated the names of the resource books they used a lot, e.g. BTI was *Bounce to it* and BEAMSP meant Beam pack *Spot the pattern*. The lists were typed onto computers so that they could easily be changed and added to at a later date. Each teacher had a copy, and there was one in the supply teacher's pack. There was plenty of space on each page for notes.

It was not always easy to say which level any activity was, especially with open-ended activities, but as the activities were more or less sequenced in order of difficulty, they were able to define some kind of border line where, say, working towards level 1 became level 1, etc. They felt that they didn't want to get bogged down in this, as so many of the tasks were differentiated by outcome anyway, and they thought that as teachers used tasks, so they would get a feeling for which child was operating at which level.

They decided to number each activity on each list, (shape 1, 2, 3, etc.) so that it could be recorded in each child's profile which activities the child had done, thus avoiding repetition in later years. They decided not to do a great tick list of all the activities, as this was time consuming and not necessarily helpful. It was much better to record what the child achieved in detail rather than great detail of what had been covered.

The next task was to decide which half term each year and class would do certain maths topics (so that equipment could be shared out), and a plan of this was made and put on the staff room wall (see page 65). The resource books and packs were kept in the staff room and were signed out if they were removed.

One teacher said to me after they had been using the scheme of work for a term, 'It's terrific! My children really love maths now, and I try really hard to give them something that is what they will enjoy. It is much easier to find something at the right level for each group now ... I never realised before how you can use one open activity and everyone in the class can have a go at it.'

DISADVANTAGES
- Watch out for those resources that mean huge photocopy bills.
- You can easily lose track of who has done what and get cries of 'We did that sheet last year with Mrs Williams'.
- The quality of some of the resources available may not be up to the standard that you want in the school – if it is just tips for a Friday afternoon or things for supply teachers to use, yes, it has a place, but not as a general way of planning what children do.

ADVANTAGES
You could place some of the varied activities from a variety of sources into your scheme of work. So for number at Key Stage 1, you might build in some worksheets from the Compare Bears, or from the Manchester Metropolitan *Bounce to it*, or from a home/school maths programme. You might want to order these in line with parts of your scheme so it might look something like figure C8. (I have invented some of the activities, as the school had not completed their work, and they were still deciding what other columns they might have on each sheet).

Scheme of work for Key Stage 1: counting to 100 and beyond 100

- Counting to 100: scheme book 4, pages 12–14.
- 'Race to a hundred' game in xxxxx book.
- 'Compare Bear' activity x, page x.
- BEAM card xx.
- 'Bounce to it', page xx, activity x.
- Counting beyond 100: scheme book 4, pages 23–25.
- Calculator activity in xxxxx, page 15 (makes a good home/school sheet).
- The 'Race to the honey' game (numbers up to 100, and adding and subtracting 10).
- 'Astronaut' counting game in xxxxx, page 17 (open-ended, so applicable for starting activity for all the class – put in own numbers – use up to 1000 with able children).
- Blank of 'astronaut' counting game to use for children to make up their own game (makes a good home/school sheet).

FIGURE C8 *Scheme of work, format 7.*

SETTING UP A MATHS GAMES LIBRARY AND HOME/SCHOOL MATHS

Many schools find some kind of home/school maths can be very helpful, but you need to think of the issues that will arise for those children who might never get the help that they need at home. You might want to have some kind of system by which you can help those children in school.

Case studies

SCHOOL A

A small school trying to build up a relationship with parents.

'We made sure that everyone had all the apparatus that they needed last year, and this year I have concentrated on getting lots of book resources for teachers to use and on getting games for the children to take home.

   Many of the teachers have used the scheme a bit slavishly over the years, and so I wanted to help them to move away from it. They weren't too sure about that, but I was as encouraging as I could be, and when I showed them some of the other resources, they really got keen on them. We had a staff meeting just looking at stuff at the local maths centre, and the head gave us a budget – quite a lot actually – and the parents gave us another £100 after the jumble sale and disco.

   We have used the books for ideas and we all have games in our classrooms that children can take home. We swap games across the same year groups, but not out of the year, otherwise we might get the 'I did that with Mrs So-and-so' when they move on.'

SCHOOL B

A large, inner-city school with a high proportion of children who do not speak English at home, and many one-parent families.

'We set up a maths games library. It took us a year to get it going properly, but it was great fun. First we had a parents' group that started copying games that we begged and borrowed from my friend's school. The parents were quite happy to copy things, and we used gift wrapping paper for the pictures or traced them. We had an INSET evening from someone who came with loads of wrapping paper and ideas and about thirty parents came. We ran a crèche for the children, and out of that evening, I should think came over sixty games. The parents finished them off for us over the next term.

Then we did an appeal on a letter to home for games that were finished with. We got dozens, and some parents went to car boot sales and jumble sales and bought things for two pence! Some of these needed new bits made, but we became quite inventive about that.

We started using the games in classes, and parents came in and played them with the children. They really liked playing the games they had made. They said that they began to feel more positive about maths themselves, and I'm sure that will help their children.

Then we spent some money on buying some manufactured games. They were part of a scheme, but that didn't matter because they could be used independently. The parents put all these into those plastic zipped wallets, and in the end we did that with all the games. Then each game had a card made for it so that it could be borrowed, and each wallet had a sticker on it saying what it contained, e.g. two six-sided dice, twelve counters, two boards and a set of rules. Then our real stalwart group of parents agreed to run the maths games library each week. What we do is that each class gets a box of games each half term on a Friday afternoon. Each child selects a game to take home and a parent signs that game out for that child. They can keep the game for a week (or two, if they want) and then they bring it back, and our wonderful parents check that everything is in the returned wallet. We couldn't possible operate the scheme without this help. They chase children for weeks saying that the Lego people from the house game are missing. We don't let a child have another game if bits go missing unless we are convinced it is a case of it being genuinely lost. We lose dice and counters, of course, but each game has the rules stuck on the back as well as loose in the wallet, and we have copies of rules in a folder in school too. We learnt by our mistakes on that one.

It works in that I think the children really enjoy the games. We can borrow particular ones from a box for the class if they fit in with our maths topic, and everyone likes that – it supports what we do in class.'

SCHOOL C
A small village school.

'We have been doing IMPACT (home/school maths) for three years now. Children take home a sheet each Friday, and most of them bring back a sheet at least some of the time. The parent and child comments on the sheets go in a small notebook, so we get a view over the term of who has done what.

I repeat some of the activities (or adapt them) in class when I can for those children who don't get help from parents. It has made our parents very interested in maths, and they often say how they wish their maths at school had been like this.'

CALCULATORS AND COMPUTERS

Some parents and teachers believe that the National Curriculum is against calculators because of incorrect media coverage, and some believe that calculators have a negative influence on children's abilities to do arithmetic.

For our children, their adult life will involve using the microchip in some form or other. We cannot escape the fact that many things, from bank accounts to washing machines, are now controlled by computer technology. It therefore makes sense for our children to use computer technology in their school lives whenever we can.

This does not mean, though, that children will not need to know things like number bonds and tables facts. What it does mean is that children will need a sound basis of mathematical understanding. This is not achieved by children just being told 'how' to do some calculation (see page 43). It is achieved by children and teachers thinking about their own methods of calculating and understanding the mathematical principles behind what they are doing.

In the CAN (Calculator Aware Number) Project (part of the Primary Initiatives in Mathematical Eduction, PrIME Project), where children, many of them infants, were given a calculator and not taught standard algorithms, the single most obvious thing about the children was their confidence with maths and their ability to work independently and in their own ways. The use of the calculator did not produce groups of children who were not numerate! The very reverse.

Of course, there are times when children should not use calculators. For example, in mental maths, a five-year-old would not need to use one to calculate eight plus two. They might use one, though, to work out what half of fifty-two is or what a hundred lots of a thousand is. A nine-year-old would not use a calculator to work out ten lots of nine, but they might use one to work out what a quarter of 31.72 was or to explore what happens when they go on halving a number again and again. (For example, halve sixty-seven, then halve the answer to that and so on.)

What is 'sensible' use of calculators?

INSET Figure C9 shows a list I have compiled over a few years working with teachers in schools.

The sensible use of a calculator will provide possibilities for:

- playing with large numbers;
- using negative numbers;
- providing meaningful maths 'play';
- checking results (but keep this to a minimum, as it can lead to a poor use of calculators – they are much more than checking instruments);
- finding out about numbers and exploring ideas;
- encouraging mental arithmetic;
- encouraging children's own strategies;
- encouraging estimating and approximating;
- deciding which calculation is needed in the context of a game or investigation;
- providing a tool to help to access new concepts;
- encouraging problem-solving strategies;
- reinforcing number bonds;
- exploring areas they normally wouldn't, e.g. seven-year-olds working with millions;
- a great improvement in attitude to maths – enthusiasm, building up confidence and getting children hooked on numbers.

FIGURE C9 *Constructive uses for a calculator in class.*

How do we actually do calculations?

Hilary Shuard (1986) said that there are three major stages in doing a calculation. Whatever method of calculating we use, the process by which we solve number problems does not change. We used to use logs or slide rules when the calculation was too big for us, now we use calculators. The process is:

1 DECIDING ON WHAT SUMS TO DO
Identify which number operations it is necessary to use in the process of solving the problem.

2 CARRY OUT THE CALCULATION
This might be a mental calculation, or a written one, or the use of a calculator, or a combination of those. Before calculation takes place, an estimation of the approximate number is often necessary.

3 MAKING SENSE OF THE ANSWER
Judge whether the results are reasonable and decide how accurate

they need to be, and be able to round the results up or down where necessary.

Those three stages are the same whether you use a calculator or not, but the beauty of the calculator is that it frees you to be much more adventurous and actually to do much more mental working out. Estimation can be seen as a fundamental part of getting a sensible answer, and we need to develop skills of estimation whenever we can.

Developing skills of estimation

You can see from the above three stages of calculating that there is some kind of estimation going on twice in the process, first to make a round guess, and then to see if what we get is reasonable. So if a six-year-old was going to add thirty-seven and forty-one, you might ask, 'Will it be nearly fifty, or nearly sixty, or seventy, or eighty, or 100, or 200?' When they have done the calculation, you can ask 'was that what you expected?' Both of those estimations are important, and they will only develop if we incorporate estimation skills into our work. Without them, children can blindly press the calculator buttons and fail to see when they get completely the wrong answer from pressing the wrong key, or getting the decimal point in the wrong place.

The National Curriculum advocates that children are not taught one method of doing calculations, but are allowed to explore their own methods, and a calculator is important in that process for children of all ages, from the nursery onwards.

We need to provide opportunities for:

- exploring what the calculator can do;
- exploring the keys and the 'light bars' and digits;
- investigating number patterns and relationships with a calculator;
- checking results in different ways;
- playing and inventing calculator games;
- discussing discoveries;
- allowing children to record in their own way;
- discussing and sharing their own methods.

What is inappropriate use of calculators?

It has been reported in the OFSTED report of 1994 that, in many classrooms, calculators are used badly. (The press picked this up and reported OFSTED as saying that calculators should not be used in schools!). This poor use of calculators certainly rings true to my own observations of the use of calculators in some classes that I visit.

If children are doing ten 'quickies' in maths off the board left by the teacher before I call the register as the supply teacher for the day, I get very concerned by those children who sneak calculators to their desk. In one class of ten- and eleven-year-olds recently, one child was using a calculator to work out eight-four times ten and 340 subtract 150. Both of those were well within the reach of that child to do mentally, and the rest of the class were calling out to me, 'Rob is cheating, Miss.' It was clear as I explored the use of calculators with that class during the day that they used calculators only for checking their work and they never played calculator games or used them to explore what numbers could do. For children of their age, they were astonishingly bad at place value and at mental calculation, and when I asked them if they found maths fun or interesting, only one child said he thought that, and the others said that was because he 'got everything right'. I set them off to find the largest number that they possibly could get on the calculator, and the largest number that they could think of they could write down, and the smallest number that they could get on the calculator and the smallest they could write down. Then I showed one group at a time how to play some calculator games. It was a very exciting day! I firmly believe that if we let children explore numbers, they really do enjoy it and can begin to develop a 'feel' for numbers. The calculator can help us in this process. Give your children calculators!

Calculators and parents

Obviously we need to dispel any fears that parents have about calculators. They naturally want their children to be good at maths and they have an understandable fear that the calculator is working against that.

If you are able to do it, running some kind of parents' session is a good idea, as it provides a place where they can air their views and discuss the place of new technology. What I do with parents is to get them to do some activities with the calculator that raise their awareness of the calculator as a teaching tool. I have put some starting points in figure C10. Any of these activities can be done with nine- to eleven-year-olds or younger children who are used to using a calculator.)

As a follow-up to this kind of activity, you could get parents to try to identify the mathematical thinking that they did as they worked on these tasks. You could ask, 'What general strategies do you use? (e.g. trial and error, looking for patterns, estimating, rounding up or down (see figure B9 on page 39).

Parents are amazed at what they have to do to work this maths out! They see that maths isn't so much about rules (as most of us were taught), but it is about looking for patterns and finding strategies – and to their astonishment, although they are using the calculator, they have to do an enormous amount of thinking.

A Multiply three consecutive numbers to find each answer. Make:

  1 a number between 400 and 500
  2 the nearest possible number to 1000
  3 the smallest possible number over 6000
  4 an odd number
  5 a number without a 0 in it
  6 a number with two digits that are the same
  7 a number that is divisible by 6. Can every number that is divisible by 6 be the product of three consecutive numbers?

B Multiply two consecutive numbers to make an even number. Is every even number the product of two consecutive numbers?
  Multiply any two consecutive odd numbers. Is the answer always odd? Can you make every odd number as the product of two consecutive odd numbers?

C Use only 3, 4, $-$, $\times$, = to generate the numbers 1–12.

FIGURE C10 *Calculator activities*

Questions parents ask

At almost every event that I have run with parents about calculators, these questions come up.

SHOULD I GIVE MY CHILD A CALCULATOR?
Giving your child a calculator for their fifth birthday would be a gift with great potential. Many teachers who give their children calculators between the ages of five and seven are convinced that those children learn maths much more effectively and much more quickly than other children. I have certainly found that to be true, but I do understand why people are anxious – calculators can be used badly.

WILL A CALCULATOR MAKE MY CHILD LAZY?
No. The reverse seems to be true. Children who use calculators from the early years have a deep understanding and 'feel' for number. Giving children open access to them seems to make children think in much more mathematical ways. I think, from what I have observed in my classes, that this might be because

when children are freed from the worry of 'getting the right answer', they play around with numbers and use their own methods (that they had before they came to school), and this involves children in very much more maths than we would have covered at this age. For example, I found that children who use calculators a great deal understand place value (one of the most crucial areas of arithmetic – see below) much better than other children.

WILL THEY BE ABLE TO ADD UP?

Sensible teachers always encourage children to estimate their answer first. If it is 120 divided by twenty-five, the answer must be somewhere close to five because there are four lots of twenty five in a hundred, and then there is almost another twenty-five. So if my answer is not close to five, I know I have pressed the wrong button.

The outcome of all the estimating and all the mental maths that children have to do to use a calculator is that children can actually add up very well. They learn to deal with larger numbers than children without calculators, and this seems to make them quite confident at an early age.

HOW WILL A CALCULATOR HELP MY CHILD?

As well as being more confident with maths, a child using a calculator is learning to deal with machines that will invariably dominate their lives in the twenty-first century. It doesn't matter where we go in the Western world, the silicon chip is there. They are in our washing machines, videos, in the cash register at the shop, controlling the aircraft as they land, in the microwave and, of course, in computers and children's games.

Punching numbers into a machine needs to be thought about, not just done mindlessly. Encourage your child to ask each time 'is that a reasonable answer?'

Some kind of poster or display and some activities at adult level and at child level put around the room can do a great deal to set the positive tone for the meeting. Examples of children's work on the wall is also a good talking point (see figure C11).

Planning for using the calculator

The calculator used well in a classroom has enormous power. The National Curriculum and the Scottish 5–14 document set out a demanding programme of teaching number that has to include using numbers in context (problem solving, etc.) and has to include children's own methods of calculating and the use of

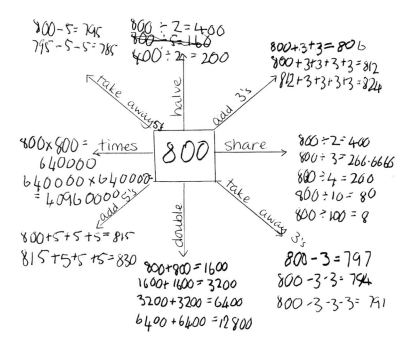

FIGURE C11   *Tariq explored 800 on a calculator.*

estimating. In other words, just routine practice of arithmetic day in, day out isn't enough.

The 1993 OFSTED report was strong on this, showing that number was not strong in schools where there was an over reliance on teacher-taught methods of calculating. Schools where children showed well developed number skills were schools with some of these characteristics.

HOW DO I SET ABOUT GETTING MY CHILDREN TO USE CALCULATORS SENSIBLY?

- Give children a chance to get used to the keys by putting a calculator in play contexts, e.g. put one in the post office with the money and the stamps, beside the shopping basket and the grocery shop, beside other counting equipment and let much older children 'play' with them when they are working on fractions or decimals or percentages or large numbers.

- Don't encourage children to check their written number work with calculators. Get them to try a different way of calculating. They will use a calculator to check whatever we do – we all do that – but minimising this use is a great help.

- Children can be helped with tables and with place value if they use calculators freely. A good starting point for using the calculator is the BEAM calculator book (see resources section).

You don't have to be a brilliant or very confident teacher to get children learning through calculators. Children love them and, given some encouragement and guidance from you about the need to estimate and work out if their answer is reasonable, they will use them to shoot off in all kinds of directions. Make a start today!

How can we use the computer most efficiently?

It is very sad that, in many schools, there is not the money or the enthusiasm to go on building on all the INSET work that was giving on computers during the 1980s. There are even some schools that do not have a functioning computer in the school and in others, the computer is used to run what I call 'arcade' programs, where children are shooting down aliens and doing such closed programs that they are learning little; the computer is used as a device for occupying children, rather than as a tool for developing children's thinking.

I cannot recommend those programs that just provide consolidation or things that can be done better in other ways. The computer is a vitally important resource in the classroom for maths, science and for developing writing skills with the word processor. My personal opinion is that you only need a good writing package, data handling packages and Logo. Things like art packages are great too, and making a newspaper is terrific, but if you want to get your children on the way to being mathematicians, one of the best ways you can do this is give them a really good long time at Logo every few weeks. By that I mean at least an hour if they are early years and can sustain it – more if they want – and not less than three or four hours – a good half a day – if they are ten or eleven. I would aim to that at least once a month with Key Stage 2 children, so there really isn't time for those programs that have monsters to shoot down or bleep when the child spells a word correctly. That is something for you to discuss at school! The child of the twenty-first century needs to be totally computer literate and able to think in a flexible and open-ended way. If they need to explore their data handling in depth, they need to know by the age of six how to enter in their data; on their own by the time they go into Key Stage 2. That means computers being used intensively for 'good' uses. (See the software list in the equipment list on page 85.)

PUBLISHED MATHS SCHEMES

It is not necessarily the answer to our problems of teaching maths just to buy a new commercial scheme! Many schools cannot do that financially anyway, so it can be a good starting point during writing the scheme of work to review the resources already available in the school. (This might be just one scheme or it might be a range of resources.)

Reviewing and evaluating resources

You could photocopy figure C12 and give it to each member of staff to fill in, either in their own time or with the school resources out in front of them during a staff meeting.

OUR PRESENT MATHS RESOURCES Review of:	Grade 1–6 1 very poor 2 poor 3 don't know 4 OK 5 does this well 6 excellent	Score
• Does what we have help to cover the present National Curriculum? • Does it help to follow up the advice in the Non-Statutory Guidance? • Does it allow for children's own ways of recording? • Does it allow for children's own ways of calculating? • Is there plenty of built-in mental maths? • Is there a balance between closed and open-ended tasks? • Are the ideas (problem solving, investigations, communicating maths, etc.) dominant in the activities? • Is there enough flexibility to allow children to develop their ideas in many different ways? • Is there a balance between the areas of maths? (Shape, movement, data handling, etc.) • Is there a balance of the different ways of learning (group work, discussion, doing practical maths, observing reflecting)? • Is there a balance between skill getting and skill using? • Is there a balance between knowledge, skills and understanding? • Are there open-ended starting points that pupils could develop over an extended period? • Are there activities and ways of working that will help pupils to enjoy what they do and that will develop them as mathematical thinkers and help them to build up a positive self-image of themselves? • Are the children involved in their own assessment?		

FIGURE C12 *Maths resource checklist.*

If you do decide that you want to buy a commercial scheme, you might want to borrow some from schools that you know or look in catalogues or ask someone for advice. If you can get down to one scheme, or two or three, you could then use this checklist to evaluate them. Pulling together all of the questions in figure C12:

- Is this scheme a significant improvement on what is currently available in school?
- Would you feel enthusiastic to introduce this scheme into school?
- Would you rather look at another scheme?

CASE STUDY

In one school, the maths co-ordinator used a sheet like the one in figure C13 and gave small groups of his colleagues a maths topic to follow through in the new maths resources they were considering buying. (So one small group followed through number and another movement, etc.) They gave each maths topic from each different scheme a score so that at the end they could see which of the possible purchases would meet their need.

You might want to use some different criteria to evaluate a scheme, and the list in figure C14 might help you to do this.

Making the best use of published maths schemes

- Used creatively, a scheme can help you to develop your scheme of work because much of the basic ground work is done for you. You might think that is a disadvantage! It depends how much you have time to do yourself and how confident you are as a teacher.
- You can use the scheme as a 'backbone'. It is there for basic structure (so in the teacher's handbook there are six activities for how to teach length to six-year-olds and what order you might want to teach some of it in).
- A scheme can be used for consolidation. Children actually like doing pages of sums, and it can help them to feel that they are achieving. Of course, at one level, a child getting the whole page right is what we want – especially if it is consolidation – but at another level, a whole page right shows us that really the child is ready to move on to the next level. It might have saved considerable time and effort if you asked the child to do a few on the page to see if they know what to do, then discuss with them what they can move on to now. Spending an hour doing thirty examples of something you already can do is a bit mindless – that's exactly why children like schemes!

EVALUATING A MATHS TOPIC			Grade 1–6

EVALUATING A MATHS TOPIC

Areas of maths: ....

Resources: ....

Grade 1–6
1 very poor
2 poor
3 don't know
4 OK
5 does this well
6 excellent

Different aspects of maths (you can add to this list)	Teacher's book	Children's book	Associated materials (games, etc.)
based on practical activities			
open-ended activities			
whole class tasks for a range of abilities			
support and extension activities			
consolidation and practice			
integrated games			
integrated assessment			
calculator/computer work			

FIGURE C13 *Chart for evaluating a maths topic.*

EVALUATING A COMMERCIAL SCHEME

Name of scheme ....

- Does it cover the present National Curriculum?
- Does it allow for teaching in line with the Non-Statutory Guidance?
- Is it flexible?
- Does it use open-ended activities?
- Does it give prominence to the use of calculators and computers?
- Does it match with the school equal opportunities policy? (e.g. how are girls portrayed in the pictures?)
- What does it cost?
- What are the 'running costs', and can you do without the consumables?
- Is it over prescriptive?
- Does it give enough support to teachers who need that?
- Does it give choices?
- Does it aim to help teachers to feel more confident with maths rather than just telling them everything that they need to do without thinking about it?
- Does it give children a 'balanced diet' of maths (see page 42)?
- Does it have a good teacher's book?
- Is it based on practical activity, even for the older children?
- Does it give prominence to the use of children's own methods or does it promote standard algorithms?
- Does it give prominence to mental maths?
- Does it look attractive?
- Does it have some home/school activities?
- Are the aims of the scheme clear?
- Are the aims and purposes of the activities clear?
- Does it focus on children discussing their maths?
- Is the assessment done in a way that we can use with our own system?
- Is the assessment built into it?
- Is the assessment about what children have achieved, or **is** it just about coverage?
- Are parents and children involved in the assessments?
- It is fun!?
- Is it too easy and undemanding for the children?
- Can you see yourself finding something for all your abilities of children within it?
- If you have children of widely different abilities, do you need many different books or are most abilities catered for in a reasonably easy-to-operate way?
- Will your parents like it?
- Can you see yourself adapting it to fit your class?
- Would you be able to control it as a scheme and use it where needed, or would it end up controlling you?

FIGURE C14 *Checklist for evaluating a maths scheme.*

'Our (scheme) maths is really easy. I like it. You don't have to think.'
(Sam, 8)

- Always start from the teacher's book.
- Don't ignore practical work and games for older children.
- Don't be afraid to miss out pages of boring practice. You could try sending it home to be done if your children love sums.
- Don't encourage anyone to let their children go from page to page and book to book. Some schools build it into their scheme of work that no child must do that.

ADVANTAGES
- The teacher's book can be helpful in giving ideas for practical activities.
- If you lack confidence with maths, it might help you to know what you need to cover and help you to learn how to do that.
- It can provide a 'backbone' to your maths teaching.

DISADVANTAGES
- It can end up controlling you.
- Used badly, it can be boring and undemanding for the children.
- A bad scheme (e.g. one where children have to fill in boxes and the maths is undemanding) will not help children to achieve their potential because there is almost no mathematical thinking involved.

Ways of supplementing a scheme

Supplementing your existing resources can be a much less expensive way of going about updating what you have than going for a whole new scheme (see page 104). Of course, you need to feel some confidence with whatever you are choosing as your 'backbone' for your scheme of work, and you might want to buy a new scheme for that, but supplementing what you already have is an effective way of putting some new zap into what goes on in the classroom.

HUMAN RESOURCES

Our 'human resources' are maybe our most valuable, but all of us know how hard it is to use our potential helpers well. I once worked in a school where parents could come in when they wanted – sometimes holding screaming infants! It drove me crazy. Although a parent popping in to see if he or she can help can sometimes be great – we have usually got something we can ask them to do – on the whole we need to be able to plan our help.

Maths tasks that can be delegated to other adults are:

- teaching and playing maths games;
- supervising the shop – are the children getting the right change?;
- supervising an activity that you have set up and have had time to give the adult some guidance on (or prepare a handout sheet), e.g. a computer or floor robot task. A parent can help five- and six-year-olds to enter their data on a data base and can also work with the children gathering the data in the first place, e.g. measuring children's weights, heights, etc.;
- working with a group of children who need extra support – either children struggling with something or a group of bright children with some very challenging work (you may well find you need to 'train' parents not to 'do it for them');
- any activity that needs the child to talk about what they are doing. Again, there is the need to say that it is the children that do the talking and that the adult is there to listen and suggest, not take over;
- organisational things, such as whose turn is it to take which game home and organising loose worksheets into the child's work folder, sorting and tidying the maths equipment with the children, etc;
- making games;
- sitting beside the water bath/sand tray/construction and discussing with the children;
- taking children to the shop to buy cooking ingredients;
- helping to make a maths trail in the school and taking children around it;
- working with groups on problem solving activities, e.g. if the children are planning a sports day, an adult can be enlisted to help with making refreshments (especially with dangerous things like boiling kettles), marking out the running track, organising the groups of children who are waiting their turn to race and anything else that the children think is appropriate for an adult to do for them.

Our human resources are our most valuable asset. Cherish and nurture them!

1   You might want to start your INSET with something reasonably simple, such as looking at your resources.
    • What apparatus do we already have? (You can use figure C12 to review resources that you already have in school.)
    • How can we share what we have?

2   How could we make the best use of what we already have? Write notes on this individually and take them to a staff meeting.
    • What maths apparatus or other resources do we still need?
    • Try to find some way to be able to evaluate other resources and use figure C13 to focus on what you need. (You could send for inspection packs of resources or visit other schools or go to a maths centre, if you have one near you.)

3   Prioritising needs.
    • *High priority*. What do we need to add to our resources in order to fulfil our statutory requirements to cover the curriculum?
    • *Medium priority*. What would help us to put together a balanced scheme of work?
    • *Low priority*. What would we really like to have but will have to wait for other, more essential things to be purchased first?

4   You might want to think of having a parents' meeting.
    • Do we need to think of having a parents' meeting to discuss some issues about maths?
    • Could we try to build up our games resources, asking parents to help us?

CHAPTER 9

ASSESSMENT AND RECORD KEEPING

This chapter includes:
1   Marking work
2   Learning from children's errors
3   Meaningful assessment practice (Shirley Clarke)

MARKING WORK

Many schools have incorporated into their policy for teaching maths some kind of system of marking children's work. The days are gone when the teachers of the older children staggered home with piles of maths books in their arms every day! But we need to have some system that helps us to know the progress of each child and helps parents to see how well their child is doing.

Children marking their own work

This works well, especially when marking is done in pairs, first seeing if their answers agree, then checking the work in some way to see if the answers are right. There are a number of ways you can teach children to do this, but not checking with the calculator every time! Reducing the calculator to a checking instrument is a poor use of such a powerful tool. You could:

- ask children to check their work using a different method from the one that they did first time, so if they used Dienes, they could use an abacus; if they did a subtraction by adding on, they could do it by some other method, such as counting back on a number line;
- suggest that if two or three of them agree on an answer and then they check it another way, they are likely to be right;
- use an answer book if you want to after children have been through an early checking system. (This is not recommended for use all the time.)

All this checking encourages children to think about their work. If we take books from them to mark on our own, they miss out on that thinking.

Marking work with parents

Where parents come into school frequently, it is possible to involve them in checking their own children's work when that work has been done for consolidation. This is very different from the kind of home/school maths schemes that can operate and can be set up where schools feel that they would value this kind of help from their parents. This way of marking can work well in conjunction with a 'maths diary' (see figures C15 and C16) that operates for all children and that any adult who works with the child would contribute to (parent, class helper, head teacher as well as the teacher).

This kind of openness about children's work may well not work in some schools, as there would need to be whole school agreement to implement such a scheme, but after several years of working alongside parents, this sharing of the marking load is a natural development for some.

Marking work with children

Many teachers will only mark children's work if the child is actually there. Combined with children first checking their work in pairs, this can be a useful way to keep tabs on what is going on. However, with enthusiastic ten- and eleven-year-olds, it could be a huge time commitment (especially at the stage when you are getting into revision for moving onto the next school or for Key Stage 2 testing), so it is probably best to have some other fall-back system.

WE DON'T WANT TO PUT TICKS AND CROSSES, BUT WHAT DO WE DO INSTEAD?

Again, the ways in which you and the others involved in marking choose to write on children's work will depend on what you decide as a whole school. Some teachers use a dot where something needs to be looked at again. Others underline in pencil, and most have abandoned red pens!

Consider that it is sometimes more helpful to make a comment rather than put ticks or crosses, for example 'Beth, you seem to understand these, well done. Number 5 needs to be looked at again. Why don't you try working with much larger numbers now?' 'What could you do to extend this idea?' This all forms a part of the 'diary' or profile, and the child can be engaged in a dialogue with you.

FIGURE C15 *A maths diary can take many forms. Two are given here; the first done in a notebook, in which the child, teacher, parent and any other adult working with the child put in notes and comments on the child's work; the second is integrated into a scheme and is done at the end of each section of work.*

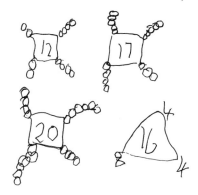

21 March   Dan did the ten-strip activity.
           Counted 10, 20, 30, 40, 50.

23 March   Played the shopping game.
           Able to identify coins.

24 March   Calculator activity

27 March   Played the 'race to
           a hundred game'. More
           confident with larger
           numbers.

FIGURE C16 *Children can be involved in their own assessment with a diary page to review work just completed. This is reproduced here with kind permission of Cambridge University Press.*

Learning from children's errors

One of the main things to say under the heading of children's mistakes is their enormous value to us, in that we get insight into children's mental images and understandings when they make errors.

INSET   (This section can be photocopied and used as a basis for staff discussion.)

One of the points noted in the 1993 OFSTED report (page 7) was the need for teachers to spend time analysing children's mistakes and working out what needs to be done to correct them. That should be built into your classroom management systems.

Consider the examples shown in figure C17. Try first to work out what each child has done. They have kept to the same rule in all examples.

1   James (8)

$$
\begin{array}{r} 936 \\ + 782 \\ \hline 1115 \\ \hline 6\ 7 \end{array}
\qquad
\begin{array}{r} 8^{1}2^{3}\!\!\not{4} \\ - \quad 471 \\ \hline 452 \end{array}
\qquad
\begin{array}{r} 137 \\ 924 \\ \hline 1512 \\ \hline 0 \end{array}
$$

2   Alison (9)

$$
\begin{array}{r} 68 \\ + \quad 35 \\ \hline 18 \\ \hline 112 \\ \hline 1 \end{array}
\qquad
\begin{array}{r} 139 \\ + \quad 284 \\ \hline 423 \\ \hline 1\ 1 \end{array}
\qquad
\begin{array}{r} 79 \\ + \quad 33 \\ \hline 19 \\ \hline 122 \\ \hline 1 \end{array}
$$

3   Shazad (8)

FIGURE C17 *Examples of children's mathematical errors.*

$$
\begin{array}{r} 24 \\ + \quad 4 \\ \hline 64 \end{array}
\qquad
\begin{array}{r} 163 \\ + \quad 14 \\ \hline 303 \end{array}
$$

When you have tried to see their errors, read the following.

1   James (8) was left handed and had a number of orientation problems and often did mirror writing. He often works from the left.

2   Alison (9) always 'carries' the smaller digit into the next column. This strategy often results in the correct answer, especially in the easier examples from her scheme. So for much of the time, her problem does not reveal itself.

3 Shazed (8) places figures in inappropriate columns. This problem does not show itself very often because mostly in his maths he copies from a book. It is only in 'real' examples where he has to write down the calculation by himself that the problem shows.

Strategies to help these children

Discuss in small groups what you could do to help these children. In your discussion, you could come up with some ideas like these to help James.

- Focus on 'Is that a reasonable answer? If you had more than nine hundred marbles and then you bought more than seven hundred more, would you have just over a thousand, or over two thousand? Can you do a quick estimate in your head?'
- Try to encourage him to write 'sums' horizontally and see if that helps him to estimate first.
- Focus on some right-to-left place value games.
- Give lots of calculator experience, as this can often help place value concepts to develop more securely.

Try to develop a similar list for Alison and Shazad (see also the action box on page 123) and think about the following:

- What will our policy be on marking?
- How will we deal with errors in our classroom?

MEANINGFUL ASSESSMENT PRACTICE (SHIRLEY CLARKE)

Assessment and record keeping is a difficult topic for all of us. It is often hard to know what is the most helpful way to assess children, and how the information what we gather can be recorded in a way that fits in with the other things we need to do each day. We know that it is useless to record so much that the next teacher who takes the child can't see quickly what they need to know.

I think that this article by Shirley Clarke says all that we need to know.

'Since Sir Ron Dearing's Final Report of 1994, the focus for teachers' ongoing assessment has been to look for **significant achievement**, as opposed to looking for **all** aspects of achievement, which had been the previous practice.

The underpinning principles for good practice assessment are:

- the assessment process must include the child, aiming for the child to become part of the evaluation process;
- the assessment process must enhance the child's learning and the teachers' teaching;
- all assessment processes should be manageable.

Defining assessment and its purpose

Mary Jane Drummond has a definition of assessment which clearly describes the process as it takes place in the classroom. She sets it out as three crucial questions which educators must ask themselves when they consider children's learning. Those questions are:

'What is there to see?
How best can we understand what we see?
How can we put our understanding to good use?'

(M. J. Drummond, 1993)

'What is there to see?' refers to the fact that we need to be able to access children's understanding in the best possible way. We need to be constantly talking to children about their work and maximising the opportunities for them to achieve in the first place and demonstrate their achievement in the second.

'How best can we understand what we see?' is the next stage. We need to be able to create a climate in the classroom where teachers are not simply hypothesising about the reasons for children's understanding, but have as much information as possible about a child's understanding coming from the child itself. We also need to be clear about the learning intentions of every activity, so that we know what we are looking for. In addition, we need to be flexible, however, because a child's achievement is not always directly related to the aims of the actual lesson.

'How can we put our understanding to good use?' is the key factor in moving children forwards. If the teacher has answered the first two questions, then the information gathered should give clear indications as to what should be the next move in helping a child to continue to progress.

The purpose of the assessment process is to make explicit children's achievements, celebrate their achievements with them, then help them to move forward to the next goal. Without children's involvement in the assessment process, assessment becomes a judgmental activity, resulting in a one-way view of a child's achievement. Information gathered in this way has minimal use. When shared with the child, assessment information is more likely to result in a raising of standards, because the child is more focused, motivated and aware of his or her own capabilities and potential. Good assessment practice enables children to be able to fulfil their learning potential and raises self-esteem and self-confidence.

'Assessment' can sometimes be used as the term for what is, in fact, record keeping. It needs to be made clear that the assessment process is that outlined so far; a means of understanding children's understanding. Record keeping is a follow up to the assessment process, and needs to take place only when significant achievement has taken place. This will be described in some detail later. Evidence gathering is part of the follow up to assessment, and needs to be centred round the idea of a 'Record of Achievement' rather than a 'collection of evidence'. It is neither a statutory requirement nor useful to keep samples of children's work at set points in time as proof of National Curriculum attainment, whereas Records of Achievement are a motivating and useful aspect of the assessment process.

The planning, assessment and record-keeping cycle: a practical solution

For ongoing aspects of the curriculum, such as language and maths, planning is focused around the scheme of work. A typical scheme of work for maths consists of A3 sheets, set out in three columns: Programmes of Study (PoS) statements and further detail/Core scheme references/Other resources. A fourth column can be useful in setting out key questions. The PoS statements are the first point of reference, with the other columns linking specifically to them. Differentiation is ensured by the statements for the whole of the Key Stage being set out at once on the sheet. The teacher then plans a range of activities across the PoS coverage for the children in his or her class, rather than trying to cover all the statements with all children.

The teacher needs to identify the range of PoS statements which are equivalent to the range of ability in the class, and plan a succession of activities accordingly. The important thing about this is that differentiation must be ensured. It is a waste of children's time to be engaged in work which is either too easy or too difficult for them. Given the Key Stage PoS statements for the development of place value, for instance, any class from the Key Stage would have children working at all aspects of this section, so the work needs to be planned accordingly. The same principles should underpin this planning as those for topic planning, in that the teacher should be very clear of the purpose of every task.

MAKING ASSESSMENTS IN THE CLASSROOM: SETTING UP THE ASSESSMENT DIALOGUE

Once the teacher is sure of the purpose of every task, the next step is to let children into the secret. By this I mean, in words which they will understand, say why you want them to do the activity (e.g. 'I want you to play this maths game because it will help you order your numbers to 100. I also want to see how well you can take turns.') This can be said to the whole class, a group, pairs or individuals, depending on how you set children off. The important thing about this is that it takes no more time than it does for the task itself to be explained; it simply needs to become a habit on the part of the teacher. It is important that children are let into the secret for two reasons:

- Firstly, because knowing the purpose focuses the child towards a particular outcome. Very often, children have no idea why they have been asked to do something, and they can only look for clues or 'guess what's in the teacher's mind' as a means of knowing what is expected of them.
- Secondly, because they are being invited to take more control over evaluating their achievements. If the purpose is known, this is more likely to encourage the child to be weighing up the relative strengths and weaknesses of their work as they are doing it.

With children informed of the purpose of the task, the assessment agenda has been set, because, when children finish their work, or are spoken to in the middle of the task, the teacher can say, 'How do you think you have got on with ordering your numbers/taking turns?' This type of questioning invites the child to play an active part in his or her learning. Children who are used to being asked such questions readily respond, giving honest answers, because they know the purpose of the teacher's questions is to help their learning process. The answers children give often put a teacher fully in the picture about the child's level of understanding, as well as why something now appears to be understood (e.g. I understand this now/Lisa helped me with these two/I didn't want to work with Sam because I wanted to do it like this, etc.).

What I have described in this section constitutes the assessment process at its best. It describes, however, the means by which the teacher makes all his or her ongoing decisions about children's learning and what they need to do next. Most of the insights gleaned from this continuous dialogue simply inform day-to-day decisions, and it is unnecessary to record them. However, when significant achievement occurs, there is a need to recognise and record the event.

MAKING ASSESSMENTS IN THE CLASSROOM: LOOKING FOR SIGNIFICANT ACHIEVEMENT AND RECORDING IT

Record keeping must have a purpose. If a teacher is to spend time writing things down, it must be useful to both teacher and child. If record keeping is focused on children's significant achievement, it fulfils many purposes. Firstly, however, we need to look closely at what significant achievement is. Significant achievement is any leap in progress, something which, from then on, will affect everything the child does. It may be the first time a child does something (e.g. sitting still for more than five minutes), or it may be when the teacher is sure that a particular skill or concept has now been thoroughly demonstrated (e.g. if the child, in a number of contexts, shows an understanding of place value). Work with teachers has led us to believe that significant achievement seems to fall into five categories:

* physical skill (e.g. use of scissors);
* social skill (e.g. able to take turns);
* attitude development (e.g. increased confidence in problem solving);
* concept clicking (e.g. clear understanding that multiplication is repeated addition);
* process skill (e.g. able to generalise).

These are all examples of possible significant achievement in the context of mathematics. Clearly, what is significant for one child is not necessarily for another. This is a welcome departure from the style of assessment which puts a set of criteria as the basis of one's judgements, rather than the child's own development.

The more examples of significant achievement one sees, the clearer the idea becomes. If a child is a relatively slow learner, it does not mean that the child will have no significant achievement. It simply means that significance has to be redefined for that child. For instance, a child who takes six months to learn how to write their name will have a number of significant events leading up to the writing of the name (e.g. the first time she puts pen to paper, the first time she writes the initial letter of her name, etc.) Similarly, a child who always does everything perfectly needs to be given more challenging, problem-solving activities in order to demonstrate significant achievement. The context within which significant achievement can be spotted is usually the ongoing assessment dialogue, although it may be demonstrated by a product, such as a piece of writing the child has done. When significant achievements occur, they can be underplayed in a busy classroom. Children have the right to have all their significant achievements recognised, understood and recorded. Recognition consists of simply informing the child (e.g. 'Well done, that is the first time you have set your work out neatly'). Understanding why the significant event took place is a crucial part of this process. It consists of asking the child why the significance occurred. In trialling with teachers, we found that the child's answer often contradicts what the teacher saw as the reason for the significant achievement. This is an important discovery, because it shows that we must find out, from the child, why significance occurs if we are to be able to follow up the achievement with appropriate teaching strategies. One example of a piece of work brought to a course on significant achievement demonstrates the importance of finding out why the significance took place:

Ben chose for the first time to be a scribe in writing up a paired problem-solving exercise. The teacher believed that the reason this had happened was because of the context of the problem ('find the height of the school'), and her decision was to give Ben more problems to do with the school building as a way of building on this success. However, when asked on the course to go back and ask Ben why he had done this, the teacher reported that Ben said, 'It was because you put me with Matthew, and he's shy, like me.' The implications for the teacher now are considerably different. Clearly Ben is sensitive to the dominance of the child he is working with, and the teacher's way forward now is to consider his pairing more carefully, both for maths, writing and perhaps for other curriculum areas.

The child should be central to the recognition and recording of the comment. During the course of a lesson, when the significance occurs, the teacher, in a one-to-one situation, needs to make much of the event (e.g. 'Well done, Ben. You have really understood ... Tell me why this happened.').

The following list outlines the features of good, manageable, formative comments which would appear on the child's actual work, or, if it is an event with no product, on a separate piece of paper which is then slotted into the child's Record of Achievement:

- the date;
- what was significant;
- why it was significant.

An example of a comment for significant achievement might be:

Ben chose today, for the first time, to be the scribe in writing up some problem solving. This is a social skill and attitude development. Ben said he was able to do this because he was working with Matthew, who is quiet, like him.

A typical child's work would have traditional comments on most of the pages (e.g. Well done, Ben) and occasional comments about the child whenever significant achievement has occurred. The formative comment has many benefits:

- The child owns the comment and has witnessed it being written, having been asked to say why the significance took place.
- Parents and other interested parties find it much more meaningful to focus on the times when a significant, formative comment has been written, because they make the progression of the child explicit.
- The child and teacher can look back to previous comments at any time, to compare with further progress and to help know what needs to be targeted for the future.

THE RECORD OF ACHIEVEMENT

The Record of Achievement is the place where any notable work is placed. The work should be negotiated between the teacher and the child. Unlike previous 'evidence collections', there is no systematic approach when using a Record of Achievement. It is simply an ongoing collection of any special work done by the child, and has been proved to be a highly motivating aspect of assessment. The Record can contain work or other measures of achievement from both inside and outside the school.

There are two places, then, where comments are written and placed when significant achievement takes place.

- Written work or other products have the comment written onto them and stay in the child's tray, or, if the teacher wants, are photocopied and placed in the child's Record of Achievement.
- Non-product style significant achievement (e.g. holding a pencil properly) has the comment written on a piece of paper (often with a decorated border) which is then placed in the child's Record of Achievement.

These records should be accessible to the child, not locked away and owned by the teacher. The ideal system is to have concertina folders in a box in the classroom. Later on, I will describe what happens to the Record at the end of each year.

THE SUMMATIVE TRACKING SYSTEM

So far, I have described the process of assessment and the accompanying formative record keeping. However, so that the system is rigorous and children do not fall through the net, there needs to be some kind of summative tracking system. This should not be a burdensome task, so I suggest the following, simple mechanism: each half term the teacher takes an A3 sheet of centimetre squared paper and writes the children's names down the side and the contexts in which significant achievement might occur along the top. These would be, essentially, the teaching contexts, (for example: reading/writing/number/shape and space (this half term's coverage)/science topic). The headings could also include the foundation subjects, but the statutory requirement is that records of some kind must be kept for the core subjects only. Bearing in mind our definition of significant achievement, however, it would seem appropriate to include all the teaching contexts, or perhaps have a further heading which simply says 'Other contexts'. A teacher in the early years would probably have different headings, such as Play/Role Play/Sand and Water/Constructional Play etc. Then, when significant achievement occurs, and the teacher has written the brief formative comment, he or she keeps track of this by entering the date and a code to show which category of significance occurred.

This tracking record can serve a number of functions. At a glance the teacher can see any of the following possibilities:

- a few children who appear to have had no significant achievement, and therefore need to be focused on, in case they have been missed because they are quiet;
- a child who has had significant achievement in say, reading, but not in writing, and therefore needs to be checked;
- the fact that none of the children have had any significant achievement in, say, science, which indicates a need for the teacher to rethink the curriculum on offer;
- a bright child who appears to have had no significant achievement, which indicates that he or she needs to be given more challenging, open-ended tasks.

END-OF-YEAR RECORDS

Anything passed on to the next teacher needs to be useful to that teacher and able to be read quickly and easily. It is of no use passing on the whole Record of Achievement, because much of the contents would have served their purpose and been surpassed by subsequent pieces of work. The best practice, therefore, is to sift the contents down to the last four pieces of significant work, say, one story, one account, one maths investigation and one science investigation. This will be manageable and useful for the next teacher to read. In the case of children with particular learning difficulties, it may be useful to pass on more pieces, perhaps showing the progression across the year. Teachers involved in the trialling of this system said that it would be unnecessary to pass on the summative tracking matrix, because it is essentially a working document.

I have outlined a framework for assessment which would first and foremost put the child's learning and development first. However, this system would also meet the statutory requirements.'

Shirley Clarke
December 1994

---

1 Do the activity on page 114 about children's errors.

2 Bring to the next meeting at least one example of a child's error. It doesn't need to be a written error. Think about the strategies that you might use to help the child to share with others in discussion. (You will need an OHP or flip chart for this activity.)

3 As you work together, you might find that you are going through a process something like the one in figure C18.
   • Where are we now?
   • How can we move on to the next stage?
   • Does anyone need help to move on to the next stage?
   • How can we provide that help?

4 Use one of the action plan sheets from appendices 2–5 for individuals to record and reflect on their responses.

5 Bring three examples of significant achievement in your class to the next staff meeting to discuss. Make notes on why it was significant for that child.

Work together to decide on how as a school you can plan to record significant achievement.

Preparing for action

Can be:
co-ordinator on his or her own;
co-ordinator and head;
co-ordinator and whole staff.

action

1   Do a thorough review of what is going on now.

2   Collect evidence – inspection reports, staff views, resources, assessment and record keeping, equal opportunities issues, liaison issues.

Action

Can involve parents and governors.

3   Identify a focus – you can't do it all! Use a questionnaire (see page 130).

4   Do a timetable (be realistic).
    – Make aims and objectives clear.
    – Have agreed deadlines.
    – How will we evaluate? Decide on criteria for success.

Implementation

You can work as a whole group, small groups, pairs, etc., but keep meeting to review what has been done.

5   Some kind of whole school agreed implementation and then some whole staff discussion of the outcomes. (Very important that this is done before the scheme of work is finalised; this might need to happen more than once.)

6   Adjust what has been done in the light of discussions.

Review and evaluation

Whole group, small groups, head with co-ordinator, governors and parents.

7   Identify strengths and weaknesses.

8   How do we improve?

(Back to step 5 if you have time and there are improvements to make before you can move on.)

9   Decide on how notes for the next review will be kept and a possible date for that review in the light of the school development plan.

FIGURE C18 *Words into action: an overview of the possible sequence of stages to write your scheme of work.*

WHAT CAN GO WRONG

- People will be away at crucial moments.
- The promised money might dry up because there is a long-term sickness amongst the staff.
- A new report/paper/media story will divert us onto something that is thought to be more important than maths.
- The governors have a crisis.
- The government will change the rules – again!
- You just run out of time and energy.

FOR THE MATHS CO-ORDINATOR AND HEAD

- You might hate the scheme and not want to use it yourself, but others might want to keep it.
- Keeping a scheme as support can be important as it might well be the key thing that will support the actual implementation of the scheme of work.
- Some teachers need more support than others, and if you take away the prop, they might collapse!
- You could use one of the action plan sheets in the appendices to plan and reflect on your next actions.

SECTION D

GETTING STARTED ON THE SCHEME OF WORK

This section is divided into four chapters:

CHAPTER 10
Towards an action plan

CHAPTER 11
Improving the profile of maths in the school

CHAPTER 12
Implementing the scheme of work in my classroom

CHAPTER 13
Monitoring the scheme of work

CHAPTER 10 · TOWARDS AN ACTION PLAN

> This chapter includes:
> 1 Working on our own
> 2 Considering the format
> 3 How is maths taught at the moment?
> 4 Making a 'worry list'
> 5 A whole school action plan

WORKING ON OUR OWN

How you can get started might depend on whether you are able to say in your situation 'how shall *we* get started?' or whether you need to say 'how shall *I* get started?'

It is probably generally true for all of us that any change, innovation or move towards making things better in our school has to start with us making some change for ourselves. That is actually all we have control over!

Even if we are maths co-ordinator or head, we need to start with something that we can do on our own, such as:

- making a maths investigation board and inviting children from all over the school to come and put up their work;
- making an interactive display in a corridor for anyone to use (you could put out some simple mathematical puzzles or some mazes or a selection of colourful maths books (see resources));
- planning a maths morning and inviting some parents to come in and help, then showing colleagues what you all did;
- planning a session of strategy games (see resources);
- planning a maths assembly such as:
  - children acting out *Jim and the beanstalk* (Raymond Briggs) with huge props of teeth and spectacles, etc.;
  - singing some mathematical rhymes (see resources);
  - an 'enormous numbers' assembly based on the numbers children have found in a project about space (a quarter of a million miles to the moon, etc.);

- running a parent and teacher workshop on making maths games. You could run a crèche and get people into teams to get games started and then these can be finished over the next few weeks. Maybe they could form the basis of a maths games library that children borrow from to take home?

If you are in a position where you are able to say how can *we* get started, you will need to do some things on your own, but you might also be able to do some things together with one of or all your colleagues. These types of collaborative actions can often be very important because it could mean that everyone (or more realistically, most people) could end up feeling that what they did was wanted, needed and partly instigated by them. People sometimes talk in terms of 'ownership'. That idea of shared responsibility is a powerful idea.

CONSIDERING THE FORMAT

You need to give very careful consideration to the question of which format you want for your scheme of work. You could:

- look at several examples from other schools or this book (see below, page 145 – 59);
- trial some of them over a term or two;
- improve formats that work for you by considering exactly what your individual needs are and the unique needs and demands of your school;
- use the list below to do an INSET activity 'A scheme of work should...' to clarify those needs.

INSET Building on the latter point, ask colleagues on their own or in small groups to complete the sentence to make clear what they think a scheme of work should do, in other words what they are expecting from this block of INSET work you are doing together.

Beware of doing this activity too early on when many colleagues may have no idea what a scheme of work should do! If you want, you could give them some starters by writing a list on a flip board or on slips of paper to give out to groups for discussion. You could use this list as a starting point and add to it things your colleagues have said. It is a good idea to include things that people will not agree with! It gets the discussion going.

A scheme of work should...

INSET We want our scheme of work to:

- be easy to use;
- be based on the xxxx scheme;
- tell us exactly what to do;
- help with continuity between schools and key stages;
- be open and flexible so that we can do our own activities;
- be in a file that we can have open on our desk;
- be a resource bank of activities;
- be clear to the parents;
- expand the curriculum in detail to show us all the steps we need to take;
- relate to year groups;
- be realistic and usable, even by supply teachers;
- relate to our school policies of equal opportunities.

It needs to have:

- clear aims and purposes;
- a clear outline structure, with the detail being filled in by teachers in their forward planning;
- clear progression;
- clear statements about exactly how we teach tables;
- enough structure to support those teachers who are less confident;
- clear activities for the less able children;
- lots of extension activities for the able children;
- clear guidance on assessment;
- bits of our scheme with other resources slotted in;
- a clear outline of the language children will use;
- lots of games and activities for all children;
- a clear statement of the resources that we should use.

You will be able to think of others that relate to your school.

HOW IS MATHS TAUGHT AT THE MOMENT?

INSET You might want to use some kind of questionnaire like the one in figure D1 for each member of staff to fill in, or these questions could be used for the basis of group discussion during a staff meeting.

You could photocopy pages 130 – 31 and ask each colleague to fill it in. It is important that this is done sensitively, so that colleagues don't feel threatened by doing it. This kind of activity is crucial in that it helps you to start your writing of a scheme of work starting where people are.

MATHS QUESTIONNAIRE

Name .... School ....

Age group taught .... Date ....

1 How do you plan your maths?

   - weekly/termly
   - on your own/in a group
   - use the scheme as a basis/relate to topic work/use programmes of study

2 How do you organise your maths?

   - children work individually/individually but within groups/pairs/small groups/whole class/mixture of these
   - If you use groups, are these static/flexible/used all the time/just used for some tasks?
   - Do you do maths within an integrated day?
   - Do you timetable maths at certain times?

3 Is most of your maths .....

   - from the scheme?
   - from other resources (specify)?
   - practical?

4 What kind of maths do you do?

   - 'do and talk' maths with discussion and activity as a focus
   - mainly arithmetic
   - mental maths – everyday/twice a week/other
   - open-ended tasks – investigations/problem solving
   - How do you balance the maths work? number/measuring/shape and space/etc.
   - Do you use maths games?
   - How do you use them?
   - How many do you have?
   - Would you like to swap games with other teachers?
   - Do you make cross-curricular links when you can?
   - How do you do that (–termly plans, etc.)?

5 How do you organise child/child and child/teacher talking? Can your children talk about their work and ask questions?

6 Do you encourage children to use their own methods of calculating? How and when do you do that? Do you teach children 'how' to do a calculation, e.g. how to subtract two three digit numbers?

7 Do you use the scheme .....

- most of the time?
- just for consolidation?
- as a basic structure?
- Do the children work from page to page?
- Do you select bits?

8 Do your children enjoy their maths?
How could you find out?
Do any of your children fear maths?

9 Do you involve children and parents in maths together?

10 How much do you use calculators and computers in maths?

11 How do you record what you do?
Do you involve children in their assessment?
Do you involve the parents in assessing the children?

12 What would you like help with?

- How to plan/use discussion more positively/make groups more flexible, etc.
- Would you like more resources?

On the basis of what is happening at the moment, through discussion and perhaps observing each other in your classrooms, you could have some maths curriculum review sessions.

- Does what we do reflect what we say we do in our maths policy document?
- Do we need to change our policy?
- Do we need to change our scheme of work?
- Can we identify some things that we could do to improve what we do? (More resources, working in ability groups across a whole year sometimes, activating parent helpers for maths games, etc.)

FIGURE D1 *How is maths taught at the moment? – a questionnaire.*

MAKING A 'WORRY LIST'

INSET This kind of list (see figure D2) can be generated by all or just a few of the staff, or just the maths co-ordinator or head. It is a list of things that people are, or might be, worried about, and a copy can be put up in the staff room for everyone to comment on, or a list can be sent to everyone.

You can write the list with five columns on the right for people to tick. The five columns might be headed: 'very worried', 'worried', 'don't know', 'no problem', 'this is going well'.

	very worried	worried	OK	no problem	this is good
1 Are we really using the scheme effectively?					
2 Mental maths is talked about a lot, but it is difficult to get started.					
3 Getting AT1 (or using applying maths,or problem solving and inquiry) is tough. Please could we have some input?					
4 What about those children who are fine with other subjects, but seem to have a 'maths block'?					
5					

FIGURE D2 *Example of a 'worry list'.*

Alternatively, you can make your list and invite everyone to mark each statement with 5 for 'very worried', 4 'worried', 3 'don't know', 2 'no problem', 1 'this is going well'. You could start work on the items that come out with a large total.

I'm glad that we...

Another way to make a 'worry list' that also highlights what you do well (and that is really important to focus on!) is to use the format, 'I'm glad that we ... (each person inserts something that is going well in their view), but I wish that we ..., (each person inserts something that they are worried about or that they think is important to discuss)'. So you might get contributions like this:

- I'm glad that we are not expected to stick rigidly to a scheme, but I wish that we had a few more resources to get ideas from.
- I'm glad that we have such a good scheme, but I wish that the parents didn't compare who was on which book and treat it all like a competition.
- I'm glad that we all have some maths resources in our rooms,

but I wish that we could find a way to store more specialist equipment in a place where we can get it easily.

- I'm glad that we do home/school maths, but I wish that we could include some more games in those activities.
- I'm glad that we have moved away from tick lists, but I wish that we could find a way that we all agree on to record children's achievements.

You might use any of the following methods to find out what your possible ways forward might be and what the priorities are:

- developing resources for more open-ended tasks;
- finding ways to build in more discussion into maths sessions;
- identifying those children with special needs (including those who are very able);
- identifying staff who need specific help, for example:
  - one teacher needs specific help using Logo;
  - another wants help to teach multiplication tables more successfully and without making the children anxious;
  - two young teachers want help to move away from the scheme;
  - the reception teacher would like more apparatus for maths and for some of the maths equipment to be stored centrally so that it is shared round more, not just sitting in the back of a cupboard.

If colleagues like the maths scheme you use, but feel they could get more from it, you could start with page 105.

A WHOLE SCHOOL ACTION PLAN

Working on maths in the school can obviously only be one part of the school development plan, and what you do first as a whole school is going to depend in part on where you are with all your other plans and what has been done so far in the maths. You will probably be somewhere in the kind of cycle shown in figure D3.

You could get started as a whole school by:

- identifying how maths is being taught at the moment (see page 130);
- identifying and listing all the maths equipment that you have in the school (see page 82);
- making a 'worry list';
- looking back to the list of what a scheme of work needs to address (see page 107).

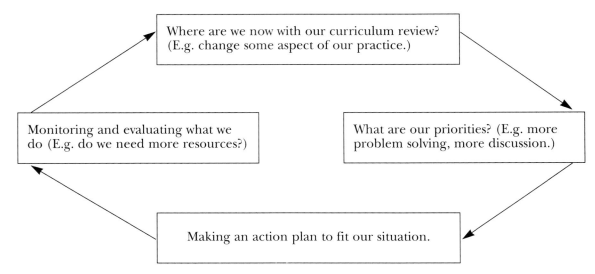

FIGURE D3 *Planning cycle.*

Formulating an action plan

It is crucial not to try to do too much. Any change needs time, and so we need to be realistic about what we could achieve in a year. Two things from this list would be more than enough, or you could maybe choose just one focus for a term:

• work on shape and space for half a term followed by a review session in the staff room, then all work on number for a half a term;
• focus just on ways to make maths in the school more open-ended and spend a year on that;
• focus for a year on children's own methods of recording and not teaching any standard algorithms;
• give every child in the school access to a calculator and see what happens after a term;
• make sure that there are the funds available for INSET and for new resources.

WHAT CAN GO WRONG?

Lots can go wrong at this stage, as probably not everyone will agree on the priorities; for example, some will be quite sure that it would solve all their problems to have a new maths scheme, and the head might have to say that there is not enough money.

If the focus is on finding what is wanted and needed by staff, children, parents and governors and matching this to the statutory requirements, you will know that at least you are heading in the right direction.

If you are having more problems than you expected, you could look back at Michael Fullan's assumptions about change on page 23 and consider whether you need to change some of your attitudes or ways of working.

Setting out a timetable

Your school development plan will dictate how much time is available, but below is the kind of timetable that has worked in some schools. Yours might be very different. What is important is to have some outline of a timetable that you try hard to keep to. You probably won't be able to, but unless you have some clearly defined timetable, you could still be working on the scheme of work in five years!

1   Starting with an INSET day.
2   Holding weekly/fortnightly staff meetings for a month/half a term.
3   Breaking up into small working groups or delegating work to a sub-group.
4   Trialling some of the formats and activities over a term.
5   Reporting back in a series of staff meetings at which sections of the scheme of work are discussed and enlarged.
6   Maybe another INSET day.
7   Having a working group to pull the ideas together.
8   Typing the work in draft and circulating it for comments.
9   Using the comments to make the final document

Remember that change is difficult and can take more time than you expect. At least one academic year is a reasonable time scale.

- Take your result of trialling a variety of formats of a scheme of work to the staff meeting. You might want to list the advantages and disadvantages of the formats you trialled.
- Do the questionnaire on page 130 about the maths in the school at the moment.
- Make a worry list using the figure D2. Share this list with others.
- Use appendix 6 to draw up a whole school action plan for the stage you are at now with your scheme of work. Make sure the timetable is reasonable.

CHAPTER 11

IMPROVING THE PROFILE OF MATHS IN THE SCHOOL

> This chapter includes:
> 1   Home/school activities
> 2   Special maths events

HOME/SCHOOL ACTIVITIES

Many schools find that considering some kind of home/school maths can raise the profile of maths. Parents often express their pleasure that the children bring home some maths as well as a reading book, often because they say they were so bad at maths themselves. Remember that:

- You don't *all* need to do home/school maths *all* of the time.
- It doesn't need to have parental involvement all of the time. Sometimes children can work independently.
- If it becomes a pressure, stop it for a term.
- Ensure that children who don't return activities are given extra help in class time.

You might want to incorporate your home/school activities into your scheme of work so a format something like the one in figure D4 could work. (See resources list for home/school maths books.)

SPECIAL MATHS EVENTS

These can raise the profile of maths in your school and they are enormous fun. They can take quite a bit of planning, but they can present maths to parents in such a positive light that just a 'maths hour' for the last hour of one day can provide discussion points for the next year! (If you have a team of parents in from lunch time, and everyone spends from then on getting classrooms and halls ready, it is quite possible to do this maths hour after just a

PROGRAMME OF STUDY THEME OR AREA OF EXPERIENCE: co-ordinates

LEVEL: _____

Learning objectives	Activities	Home/school	Other columns (e.g. language, resources, etc.)
understanding of position on a grid	1) Battleships game 2) What's in a square? 3) Pirate treasure game 4) Make a map	• Co-ordinate game • Map game	
plotting and identifying positions and recording as an ordered pair (B4, etc.)	5) Map the village	• Where is it? • Can you find the treasure?	
positioning points using Logo	6) Find the treasure 7) Grid activity		

FIGURE D4 *Scheme of work, format 8.*

few weeks of planning. A maths day or afternoon would be more complex, but don't let that put you off! Schools that do that often look back on their 'maths bonanza' as the time when maths really began to 'take off' at school.)

The idea of a special event is to get everyone doing maths and for parents to go around looking at it all. An easel or poster by each activity can explain the purposes of the activity and the maths involved (see figure D5).

You might also want to show the different tasks that the children are doing, maybe on a handout sheet or another board, or get the children to write and draw what they have done in class time that demonstrates a balance of tasks done. You might end up with a list something like this:

TASKS

- Finding out which of these containers holds the most and the least.
- Two of the plastic bottles marked with yellow tape hold the same amount of water. Which two do you think they are?

FIGURE D5 *Activity display during a 'maths day'.*

- Estimate how many yoghurt pots full of water you think the bottles marked with red tape hold. Then see how close your estimate was.
- Which bottles hold more than a litre? Which hold less? Estimate first.
- How much water do you think there is in the water bath altogether?
- Can you get all the water into the buckets only using the plastic tube?
- Does the tall litre measuring cylinder really hold the same amount as the litre jug?

PURPOSES
- To compare capacities, especially of tall, thin and wide containers.
- To begin to appreciate what a litre is.
- To learn to measure capacities.

Here are some of the activities that one school did in an afternoon:

- a cafe selling sandwiches and cakes children had made and serving tea and squash;

- all the computers in use, plus a turtle and floor robot;
- sewing patchwork and relating this to tessellating shapes on children's paper designs;
- one whole room was full of maths games;
- one group ran a 'puzzle corner';
- a maths trail around the school grounds;
- construction activities going on showing how children learn to measure and use the properties of three-dimensional shapes.

Around the school, there were exciting displays of maths done by the children ranging from some calculator work by the six-year-olds to a three-dimensional treasure island where you could pick your square using the grid reference to guess where the treasure was buried.

The key for the afternoon was definitely 'activity', and there was a parents' evening to follow at 7.30 p.m. One parent said, 'I'll never think of maths in quite the same way again'.

- What could we plan to do to improve the profile of maths in the school?
- Maybe the maths co-ordinator could work with a small group of colleagues to generate some ideas to suit your school. Then bring these ideas to a staff meeting.
- What could I do to improve the profile of maths in my classroom? Use one of the action plan sheets from the appendices to make notes on your ideas. It is helpful to think in short-, medium- and long-term plans.

CHAPTER 12

IMPLEMENTING THE SCHEME OF WORK IN MY CLASSROOM

> This chapter includes:
> 1 What does the scheme of work mean for me?
> 2 Will this format work in my classroom?

WHAT DOES THE SCHEME OF WORK MEAN FOR ME?

When all the staff meetings are over and the scheme of work is written, then the real work starts of actually doing it in the classroom. This will be a gradual process, starting from the trialling as ideas for the scheme of work are being evaluated. We must not expect that our classroom teaching will be revolutionised overnight and that our children will sit around beavering away at exceptionally brilliant maths tasks and begging to take them home at the weekend, returning with solutions that would look good in a GCSE folder. Life just isn't like that.

It is much more likely that we will be able to plan more effectively within a term or so, that we will gradually detect a growing confidence in ourselves and our children, and that we will get the hang of using open-ended questions eventually, with lots of set backs and kicking ourselves for making interventions that we wish we hadn't.

How do I plan for these children today?

First, you could just take one significant thing for you that you have identified by doing the activities in the action boxes, for example getting children to work in pairs rather than individually when they are doing a page of 'sums'. Or you could look at what you already do that you like and that works in your classroom. Plan to keep some of that, as it will give your children some security and help you not to feel that teaching is too overwhelming to cope with.

Then plan to implement just one of the aspects of your new scheme of work this week. Here are some examples I have gathered from teachers of the things that they made a start with.

- Making just one of the four maths sessions a week into an 'activity' maths session, focusing on 'do and talk' maths.
- Making all the maths for just one week based on open-ended activities (see pages 12 and 33), then evaluating that with a colleague.
- Planning to do similar activities with a colleague and then evaluating them at the end of a week or fortnight.
- Putting children in groups and letting them make their own maths game, then over the following weeks making space in the week for children to play and develop each other's games.
- Making a folder or box with free choice maths activities available when children have done their other work.
- Getting all the maths apparatus out onto open shelves, putting it out at the start of each maths session and praising children who use it.
- Planning a set of tasks using the 'balanced diet' planning sheet (see page 45), evaluating that with a colleague and, in the light of that, planning another set of tasks.
- Making a 'maths area' in the classroom.
- Buying a new and good resource book and sharing it with the children and selecting some tasks to try over a few days.
- Do an 'explode a number' (see figure D6, the PrIME folder (Shuard *et al.* 1990) pages 13 and 14, and Atkinson, 1992, pages 91–97).

You can probably think of other ways to start. What matters is that you do start, that you start with something that will fit you and your children and that you evaluate what you do, and reflect on it so that you can develop your practice.

Implementing the school scheme of work in my classroom

How you translate your school scheme of work into your classroom will depend on how detailed it is, and the form that it takes. If your scheme is very detailed, with activities described, then you need to plan just the 'how' and the 'when' and select what you will do with which children. If your scheme of work is more of an outline of what needs to be covered, you will need to plan the activities as well using the resources available to you.

For many schools, the way of actually making the scheme of work has been to get teachers in their classrooms to plan a section

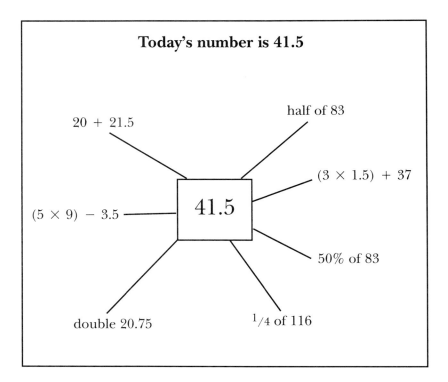

Today's number is 41.5

20 + 21.5

half of 83

(5 × 9) − 3.5 — 41.5 — (3 × 1.5) + 37

50% of 83

double 20.75

$^1/4$ of 116

FIGURE D6 *'Explode a number' can be done at any level. It is such an open-ended task that it is suitable for all abilities.*

of work for their children, then carry it out, then evaluate it and change it, then incorporate that into the scheme of work. So if one teacher plans a half term of work on multiplication with her nine- and ten-year-olds, a tidied up version of that becomes part of the scheme of work for Key Stage 2 children for the school, and others will add to it as they cover multiplication.

If you use figure D7 to look at alternative ways of getting going, and link this with format that you want to try out, you will be able to make an action plan for the next few weeks. For example, if you like a format working from your resources in school, such as the format shown in page 76, you could share out the work load (relating this to what you are going to teach this term) and maybe work in pairs to develop some key activities.

One possible way that you could work as a whole staff but spreading the workload is shown in figure A7 on page 25. You can go through this process many times before completing your scheme of work.

In the end, maybe after at least a year, various aspects of the scheme of work are done and the outline is there. Now you need to:

- check that nothing is missing from the Programmes of Study;
- tidy it up;
- type it out;

GETTING GOING

You could:

| 1 Take on **OR** aspect of the PoS at a time and spend at least a term developing key activities. | 2 Work in **OR** small groups/ pairs taking a level or age group, and develop key activities for one area of maths at a time. | 3 Write up **OR** the topics as you teach them one at a time over at least a year (better to take two). | 4 Sketch in **OR** an outline of each main area of maths and fill in the details as you reach them. | 5 Plan the whole of the maths for the school in outline so that scarce resources can be shared, and required coverage is there, then fill in the detail each term as you plan it. |

Or you can use any combination of these – or something quite different. Choose what will suit your school.

It is vital that small groups meet regularly with the whole staff group to share plans and report back on trialling.
Modify your plans as you go along, otherwise you might forget suggested changes.
Keep making reference to the ways that you do your classroom planning and to your overall school development plan.

Keep reviewing, trialling, talking and meeting.
What you end up with must work for you in your classroom.

FIGURE D7 *Ways of getting going with your scheme of work.*

- get it into each classroom, probably with a view to adding to it over the following two years and amending it at the end of that time.

WILL THIS FORMAT WORK IN MY CLASSROOM?

As will be clear from this book, there are a huge number of different ways that you can plan out your scheme of work. Some of the more popular ways are presented here, showing the detail that you need to develop for your classroom.

If you want a scheme of work in a format not shown in this book, that is absolutely fine. As long as you start with the Programmes of Study and cover the whole curriculum, the actual format is up to you. But what a scheme of work must do also is be easily translatable into the detail necessary for the youngest and least confident teacher to implement in their classroom, with all the limitations they have of numbers of children, the wide range of ability, etc. Don't leave that to the end! Work on it throughout the process, otherwise you are in danger of ending up with something that is not going to work.

Scheme of work format 9

This scheme of work has at least one activity for each 'level' for every area of maths. The example shown in figure D8 is length for Key Stage 1.

ADVANTAGES

Each aspect of the curriculum is represented in the scheme of work by one very good and reasonably open activity. Schools that use this format actually develop a much wider range of activities than just the one in the scheme of work, but even if the children just did this one activity, they would be getting a good balance of maths because each child must do every activity plus whatever else the teacher plans. (And it is open to him or her what else is planned.)

DISADVANTAGES

You might not want to focus in this tightly on levels. We don't want to reduce maths to the minimum. There are some terrible books on the market that head in this direction! (Some teachers enthuse about these and say 'it's all done for you!') There is the problem of repeating the activity in different classes (some that use this scheme have gone on to develop an activity at most levels for the different years that children could be in at that level).

MATHS measuring length	PROGRAMMES OF STUDY • comparison of lengths	
Concept Indirect comparison of length		**Notes**
Activity Would the PE shed fit through the double doors into the school? (Or choose another object that cannot be moved so that children have to measure.)		**Assessment** • language of measuring • choice of equipment • accuracy of measurement • attitude to task
Extension Would the measurement of length around each child's head in our class reach right across our room from the bookcase to the door?		**Language** wide/wider than measure narrow/narrower than
		Calculator/computer
Resources cubes string (child to select)	**Reference books**	**Cross-curricular links** Fits in with design topics, building models

FIGURE D8 *Scheme of work, format 9.*

We need to keep a careful record of which child has done which activity and with whom to avoid repetition.

HOW DO I USE THIS IN THE CLASSROOM?

The activities would be covered in a systematic way according to the area of maths being covered. Additional activities would need to be added to suit the children. So for length, a seven-year-old might include these activities as well as the one above.

- How far in paces do you think it is across the classroom? Estimate first then measure.
- Is it true that a person's reach (arms stretched out, measuring finger tip to finger tip) is the same as their height? Investigate this, putting your results on the data base.
- Can you drive the floor turtle along this roadway and into the garage (see figure D9)?

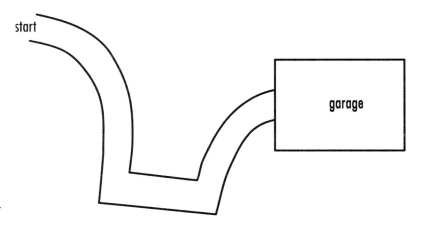

FIGURE D9 *Route for floor turtle activity.*

Scheme of work format 10

Develop a scheme of work that lists the maths to be covered. This is therefore an expanded version of the curriculum. Many schools have a scheme of work like this and figure D10 shows a section from one of those.

Many of this type of scheme of work have no activities suggested in them at all, but you can see from the example how you could add columns of activities, resources, language, using and applying maths links and anything else you think you need. If you do that, you need to think how you would manage that in the classroom. One school has A3 sheets folded and punched so that they fit an A4 file.

MATHS *number/algebra*

Year range	Number notation/place value		Resources/references
Reception →	**Number notation/place value** • Recognise number words, say rhymes, etc. • Match 1 to 1 • Count to 5/10 • Use language of number (more than/less than) • Order sets zero to 10 • Ordinal numbers 1st to 10th • Estimate up to 10 • Use half in real contexts • Write numbers 1–10		
Y1 →			
Reception →	**Operations** • Mental addition and subtraction to 5/10 • Estimating to 20 • Comparing two sets/more than/fewer than • Own recording of + and – • Using calculator for + and – • Uses +, – and = keys • Uses 'take away', 'cross out', 'subtraction' in context to 20 • Links made between + and – • Can you find missing number 7 + ... = 10		
Y1 →			
Reception →	**Algebra/pattern** • Recognises simple pattern		

FIGURE D10 *Scheme of work, format 10.*

ADVANTAGES

It is up to each teacher which resources they use for each part of the curriculum. Many like this freedom to choose their favourite activities and resources. Teachers say it helps them to be creative and to keep their maths stimulating and interesting, especially when they have children for more than one academic year.

Also, the exact requirements of the curriculum are very clearly laid out.

DISADVANTAGES

- There is the danger of repetition. (This is reduced if there is lots of choice of resources and if you put in some planning of who will use what and when.)
- There is a lot of work to do for each teacher to implement the scheme of work.
- The school needs to have a wide variety of resources to carry this out. (You might see that as an advantage!)
- Teachers must check carefully that they have covered every aspect with their chosen activities.
- With no specified activities for using and applying maths, some teachers may not have the knowledge and understanding of these aspects of maths to cover this part of the curriculum adequately. So listing using and applying aspects is as crucial as listing the content.

Scheme of work format 11

Some schools break up the maths curriculum into mini-topics that link in other curriculum areas (see figure D11).

ADVANTAGES

It is similar to the topic teaching that many teachers use, so it can link well into what is happening in the school at the moment. It also leads to a cross-curricular way of teaching that many believe helps children to make links between things, and this helps their learning.

DISADVANTAGES

It is important to make sure that all the content of maths is being covered. Ideally, this should work if the starting point is the Programme of Study, but our experience tells us that it is all too easy to get side-tracked when we teach from topics. If you were to link this format with one of the others, you might be able to be more sure of your content coverage.

MATHS PROGRAMME OF STUDY	MINI TOPIC PROGRAMME OF STUDY	CROSS-CURRICULAR LINKS
Properties of 2D shape Properties of 3D shape	Christmas (stars, boxes, decorations, printing, wrapping paper, etc.)	
Symmetry	Christmas (line symmetry in pictures, etc.) Making a kaleidoscope	art science

FIGURE D11 *Scheme of work, format 11.*

Scheme of work format 12

Figure D12 shows a sheet suited to use alongside a previous exploration of the content of maths from the Programmes of Study. (You could use the diagram on page 58 to help with that.) You can then fill in the activities, ensuring that you have covered all aspects of the curriculum.

Scheme of work format 13

If you like the idea of planning out what is to be covered over the whole year first, the chart in figure D13 could help you to see how to cover all the Programmes of Study over the year. You will probably find that you can't cover everything, so it is important that these whole year plans are done in co-operation with the teachers in the years either side of yours.

Scheme of work format 14

Starting with a maths topic (see page 76) is always a popular way to teach, and some topics may require this, as they might not blend well with the cross-curricular topics you are doing that year.

LANGUAGE

Maths Area: _____
A T: _____
Level: _____
Year Group: _____

CONTENT
(P of S)

ASSESSMENT
CRITERIA

RESOURCES

EQUIPMENT

ACTIVITIES

GROUPING   RECORDING

FIGURE D12 *Scheme of work, format 12.*

AUTUMN		SPRING		SUMMER	
NUMBER	NUMBER	NUMBER	NUMBER	NUMBER	NUMBER

DATA HANDLING →
(ongoing from topic work)

MEASURES AND SHAPE & SPACE

FIGURE D13 *Scheme of work, format 13.*

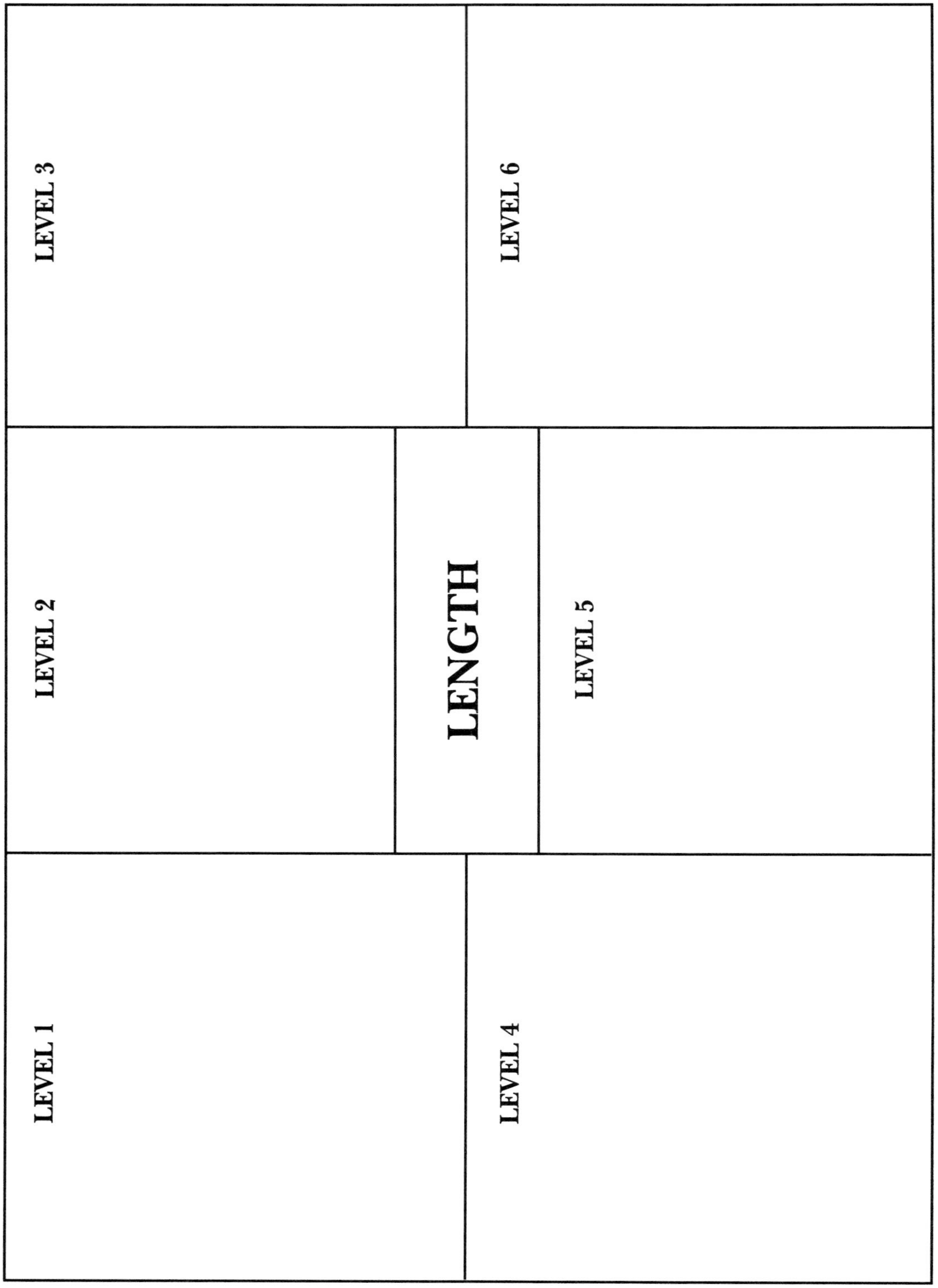

LEVEL 3	LEVEL 2	LEVEL 1
	LENGTH	
LEVEL 6	LEVEL 5	LEVEL 4

FIGURE D14 *Scheme of work, format 14.*

You might not want to focus some activities down just to one level, but when you are thinking about developing a maths topic such as length, it might help the whole school planning to see something of the progression to be made from the five-year-olds to the eleven-year-olds. You could do a chart like the one in figure D14 for each mathematical area (see page 90).

Scheme of work format 15

If you don't like the close link to levels, you could use the idea of 'early', 'intermediate' and 'advanced' activities, perhaps including some support and extension activities as well (see figure D15).

Scheme of work format 16

Many schools find that a scheme of work needs more than an A4 sheet for it to operate successfully. Figure D16 is one example that needs A3 sheets.

Scheme of work format 17

If you have found the planning sheet on page 45 successful, then you could use that sheet, or some adaptation of it, for your scheme of work. Figure D17 is a simplified version of the planning sheet in which the different teaching styles are listed down the left and the activities are filled in for each unit of work, maybe half a term on shape and space, or a fortnight on revising multiplication.

Scheme of work format 18

A very similar format is shown in figure D18. In this one, you need to try to find at least three practical activities for each unit of work – three games, three investigations, etc. Schools that use a sheet like this say it works well in translating to weekly classroom plans as well.

Scheme of work format 19

If you want to develop a scheme of work like those shown in figures C1 (page 73), C6 (page 89) and C8 (page 93), you can do that by delegating areas of maths to individual teachers or (probably better) pairs of teachers to develop. They work on their plans for classroom work over the term and then tidy up the work so that it can become a part of the school scheme of work.

EARLY

- Compare without measuring
- Order without measuring
- Free plans

EXTENSION

- Finding formulae through practical experience
- Use of IT
- Areas of circles

LANGUAGE TO USE

- short, shorter, shortest
- long, longer, longest
- more than/less than
- surface, around

Estimation and approximation used in all activities.

Measures and shape

INTERMEDIATE

- Conservation of:
  - volume
  - area
  - length
- Structured play
  - irregular/informal units
  - regular units
  - need for standard unit
- Using standard units
  - perimeter of simple shapes
  - area found by counting squares
  - volume by using cubes

ADVANCED

- More accurate measurements
- Circumference of circle from practical experience
- Use of scale
- Areas of irregular shapes

LANGUAGE TO USE

- area, volume
- perimeter, space taken up
- length
- m^2, cm^2, m^3, cm^3

FIGURE D15 *Scheme of work, format 15.*

AREA OF MATHS

LEVELS/KEY STAGE/YEAR GROUP
[This will not always be clear if you use open-ended activities which need not be too level related]

Programme of Study (and any missing links)	Assessment objective	Suggested activities	Organisational notes	AT1 notes	Language	Resources/ reference
			[This can include grouping different learning styles and teaching styles (balanced diet).]			[You will find you keep adding to this column, so make it wide.]

FIGURE D16 *Scheme of work, format 16.*
You need to draw this out on A3 paper.

Unit of work	(name of unit, i.e. topic name) date					
Practical activities						
Calculator/ computer						
Mental maths						
Written work						
Number activities						
Other						

FIGURE D17 *Scheme of work, format 17.*

MATHS TOPIC: _____   **LEVEL:** _____

YEAR GROUP: _____

Aspect of maths (balanced diet)	Activity name	Programme of Study	Notes
Practical activity	1) 2) 3)		
Games	1) 2) 3)		
Investigations	1) 2) 3)		
Calculator, computer, floor robot	1) 2) 3)		
Scheme of work and consolidation	1) 2) 3)		

FIGURE D18 *Scheme of work, format 18.*

ADVANTAGES
- You can do the bulk of the work as you do your half term plans.
- As you then carry out the plans, you can adapt what you initially wrote down and make it better for next time.
- You can choose to take the same area of maths throughout the school and thus build up a complete list across the levels.

DISADVANTAGES
- Takes a long time to produce.
- Can be quicker if different sub-groups take a different area of maths and the co-ordinator pulls it together. That is an incredibly demanding task, and you need to give at least a year. The point that a scheme of work is an ongoing and dynamic document has real meaning here.

Scheme of work format 20

The last format given here (figure D19) is a scheme of work that you can buy. It is called *Mathshare* and is sold in a ring binder with the left-hand page describing the Programmes of Study and the right-hand page blank for your notes (references to your school resources, assessment pointers, suggested groupings for activities, etc.) so that you can make it work for your school.

If you want something to get you started which could be the core of your scheme of work, you can buy *Mathshare* from Muriel Chester (address in the resources). Many schools have used it and find it valuable because it is clear, easy to use and with plenty of space for you to develop your own ideas and activities for your class.

Remember:

- However we plan, it is a great help to use open-ended maths activities as starting points because these can often be used with all of the children who will do it at the level that they are capable of – often surprising us!
- It is because children so often can do more than we think that it is so essential to use open-ended activities.
- If we give children what they *think* they can do, we are often limiting them and not letting them spread their wings and really achieve their potential.

SHAPE AND SPACE : ANGLE

Pupils should understand that angle is a measure of turn and use it in the study of shape, rotation and direction

LEVEL	PROGRAMME OF STUDY	LEARNING EXPERIENCES
4	• Constructing simple 2D and 3D shapes from given information and knowing associated language	• Construct squares, rectangles, other quadrilaterals, pentagons, hexagons, etc. using geostrips. Push out of shape and note any angle changes, e.g. from right-angle to acute or obtruse, from acute to reflex angle. Make shapes on pinboards – identify and measure acute, obtuse and reflex angles. Use a ruler, setsquare, protractor to draw a variety of shapes from given information about their angles and edges. Investigate the sum of the angles in a triangle; a quadrilateral. Understand the meaning of interior and exterior angle. Draw a range of shapes using LOGO and realise the important of exterior angle turns in their construction. Experience the angle turns by physically walking the shapes before using LOGO. Simple tessellations – investigate angles at the junctions of fitting shapes. Why do some shapes fit exactly and others don't?
	• Making sensible estimates of angles	• Draw a rough sketch of an angle given its size. Given a range of drawn angles, estimate their size to the nearest 5°. Estimate any equal angles. Check using a rotagram and a protractor. Useful software: ANGLE 90 (SMILE 31) ANGLE 360 (SMILE 17).
5	• Measuring and drawing angles to the nearest degree	• Use both the semi-circular and circular protractor to measure angles to the nearest degree. Construct a wide range of regular and non-regular shapes from given information about their angles and edges: (i) practical work with instruments, (ii) using LOGO. Establish that the angle sum of a triangle is 180° and of a quadrilateral is 360°. Measure the angles of shapes at the junctions in tessellations and tiling patterns. Understand that the angles at a point total 360° – relate this to tessellations of equilateral triangles, regular hexagons, non-regular triangles and quadrilaterals. Understand that angles adjacent on a straight line total 180°. Investigate exterior and interior angles. Draw regular polygons using exterior angles (with instruments and using LOGO). Be aware of everyday uses of angle – ramp, wedge, gradients, elevation, surveying, bearings, longitude and latitude, snooker. Angles work in height finding, using shadows or a clinometer. Use of a theodolite in surveying. Understand the convention that bearings are always measured clockwise from North and that N, E, S, W have bearings 0°, 90°, 180° and 270°. Explore (i) bearings in the classroom from a central point, (ii) bearings in the local area with the school at the centre, (iii) bearings of major European cities from London as the centre. Understand a radar map and use it to locate position. Useful software: SNOOKER (SMILE 31).
	• Explaining and using properties association with intersecting and parallel lines and triangles, and knowing associated language	• Investigate the relationships between angles made by two intersecting lines, e.g. diagonals in a square, rectangle, other quadrilaterals. Find pairs of equal angles. Similarly investigate angles made by a line crossing a set of parallel lines – find sets of equal angles. Explore the angle relationships in various capital letters.

FIGURE D19 *Scheme of work, format 19.*
Thanks to Muriel Chester for permission to reproduce this page from Mathshare.

- Look at the different schemes of work examples given here. Look through the advantages and disadvantages and make your own brief notes on the format you would like most and why. Take your notes to a staff meeting.
- Decide on some formats to trial in your class.

CHAPTER

MONITORING THE SCHEME OF WORK

This chapter includes:
1 Diversity and equality
2 The maths covered
3 Evaluating outcomes
4 Developing the scheme of work as a dynamic resource

DIVERSITY AND EQUALITY

 INSET There will be some unique aspects of our schools that will require some thought from the outset. You could generate a list of these as a whole group. Examples might include:

- Provision for children with disabilities. (Do we want to recognise some kind of maths 'block' or mathematical dyslexia?)

- Provision for children with other special needs.
- How can we be sure we are giving children who do not speak English at home a good chance to excel at maths?
- Do we have lower expectations of some children because of their race?
- Do we have lower expectations of some children because of their class or gender?
- How can we help girls to develop their potential in maths? (Boys, too, but some people believe that girls are 'naturally' worse at maths than boys.)
- If we have a home/school maths project, will it be divisive?
- How can we best use the school environment?
- We desperately need more computers/floor robots/concept keyboards/bucket scales/etc. How can we manage that?
- How do we plan the best use of the resources that we already have?

- Do we have attitudes and expectations about maths that in any way hinder children's progress?
- Use of classroom helpers and parent help.
- If we make the use of language important in our activities, how do we allow for those 'silent' children to keep the low profile they clearly want?
- How are we going to address the needs of those children who excel at maths?
- How will we attempt to make our scheme of work reflect our multicultural policy?

THE MATHS COVERED

It is part of the co-ordinator's role to monitor the maths throughout the whole school, but that is a huge task without some agreed strategies for him or her to do that.

- Time to work alongside others.
- Time set aside for regular reviews.
- Discussions with the head on ways forward in supporting individual staff.
- Reviewing the content of maths taught. Is the whole school yearly plan realistic and is the content set out there actually being covered?

A well-planned scheme of work can mean that it is possible to keep an eye on the continuity of work, but this is another issue that needs to be reviewed, perhaps as a part of senior management discussions and during liaison meetings.

The progress of individual children who are a cause for concern is another aspect of the co-ordinator role that can only work if there is a whole school discussion on how that can be done. (Don't forget to include children who shine at maths).

EVALUATING OUTCOMES

It is tempting, once the scheme of work is written, to get so bogged down with the next, vital part of the school development plan that making notes for needed changes to the scheme of work can easily get squeezed out in hectic day-to-day life. Schools who leave spaces for notes are usually very glad that they did that, as jottings can be made at the time you are using the scheme. Similarly, an A4 ring-back folder can easily have pages slotted in for extra activities and notes, so that when the next review comes around, most of the work is already there.

Part of our evaluation will happen as we assess the outcomes of children's work and report to the parents and governors. It is hard to find the time, but there are questions that we need to ask about both our work and the children's.

- Is there evidence that I really am covering all the curriculum, including using and applying maths?
- Are we raising standards?

The latter is a very hard question to answer, but we need to try to evaluate whether our system of planning and delivering our scheme of work is leading to children demonstrating a 'feel' for a number and a confidence that arises from good teaching. You might want to think of ways of collecting this kind of evidence. Anecdotes make a good start, as do examples of children's work that demonstrate children thinking for themselves.

Self-evaluation

However much we do together as staff, nothing is quite so effective in improving the 'delivery' of the curriculum as each one of us asking ourselves questions about our practice.

AM I ACTUALLY DOING WHAT I THINK OR SAY THAT I AM DOING IN MY PLANNING?

That's another tough one, but unless we address that in some way, there could still be gaps and weaknesses that might be quite easy to remedy. Working with just one colleague that you really trust can help if you can take time to observe each other and meet informally for a chat over a cup of tea maybe each month. (Informal chats can be very significant in our professional growth, but unless we set aside a specific time and put it in our diaries, it is another thing that will get lost in the busy, everyday life of a primary school.) I don't think it should be just new teachers who have mentors. I think we all need a 'buddy' if we are going to develop professionally.

If you want to look at what you do in some critical way, the Open University '*Curriculum in action*' pack has six questions that have been used with many teachers to analyse their work.

1   What did the pupils actually do?
2   What were they learning?
3   How worthwhile was it?
4   What did I do?
5   What did I learn?

6   What do I intend to do now?

(These questions are from *Curriculum in action: an approach to evaluation*, Open University.)

You can put them on your desk and take just a few moments out of a session of teaching to make quick jottings as answers for your classroom at that moment. Teachers who really make an attempt to observe themselves very closely in order to improve what they do, say that those questions are sometimes the start of them being teachers who just 'coped' to being 'reflective practitioners'. Working on these questions with a colleague is ideal, but of course you can do it on your own.

DEVELOPING THE SCHEME OF WORK AS A DYNAMIC RESOURCE

Just being involved in the putting together of a scheme of work can be hugely influential in our professional development. However much we might think that we stood on the side-lines, the talking, thinking, trialling and deciding will have influenced our thinking in some way.

We can take steps to accelerate our professional development in maths by:

* observing ourselves as described above;
* identifying our strengths and weaknesses and making some kind of action plan to fit them;
* discussing what we are doing with someone;
* building a section on our maths teaching into our yearly/termly appraisal of our work.

Developing an action plan for professional development

* Have small goals that you try to attain.
* Set a long-term goal: 'Where would I like to be in a year?'
* Build in time to reflect on what you do. (Keeping a reflective diary is ideal, but just a few jottings every now and then can greatly help your reflections.)
* Ask for support that you think you need.
* Arrange a visit to school that teaches maths in a different way from your school – this can be enormously helpful.
* Ask the maths co-ordinator or head to talk through with you what you could do.

Looking outside school for support

There are a number of courses available in maths from many institutions. Some of these use distance learning, so it doesn't matter where you live, you can still have access to a course.

- Write to your nearest college of higher education or university.
- Write to the Open University (see resources).
- Ask an advisory teacher for ideas.

There are a number of organisations that can support you, and it is worth asking around before you commit yourself to something. The Mathematical Association and the Association of Teacher of Mathematics (addresses in the resources) are always willing to help, and there might be a regional group in your area. Ring them up in office hours and find out.

Just going to a bookshop, a library or a county resource centre (if you have one) and browsing for a while can be a source of inspiration. We *need* inspiration. It is hard to go on improving what we do without it.

A last word

I asked some of the teachers that I have worked with while I was writing this book for their advice to others who are still plodding on. They said:

- Don't give up!
- Be realistic.
- It isn't the end of the world if the inspectors come and the scheme of work is only partly done.
- Keep trialling.
- Keep saying 'is this going to work for me with my children in my classroom?'
- Be kind to each other.
- Don't be afraid to admit your mistakes and your fears.
- If you keep going, you will find that you are much more confident about teaching maths because of all the hard work you did together.

- Individually use one of the action sheets (appendices 2 to 6) to identify some things to work on this week.
- Discuss the factors that influence number skills as they are described by OFSTED (see figure D20).

Factors that did not support the secure development of number skills in pupils

- Excessive practice and the overuse of number workbooks and worksheets.
- An insistence on standard algorithms for computations.
- An over-reliance of pupils on teachers showing them the appropriate method.
- Teachers giving insufficient time to finding out why children had made mistakes.
- Over-reliance on a single strategy for working for all children.

Factors that seemed to encourage children to develop good number skills

- Placing number work within a context that had relevance and meaning to the children.
- Teachers being given time and support to develop skills of analysing pupils' mistakes, and to observe each other in the classroom.
- Monitoring number work throughout the school to maintain consistency and quality.
- Ensuring a balance in the children's maths between mental, practical number investigations and practising skills (consolidation).
- Children developing a feel for number through interesting and challenging number investigations.
- Ensuring that number work is developed across the curriculum.
- Emphasis being given to the language of number.
- The everyday use of calculators and computers as part of maths work.

FIGURE D20 *Factors that influence number skills (OFSTED).*

For the maths co-ordinator and head

Using one of the action plan sheets in the appendix, you could spend some time on your own writing notes on some of these items.

- What could I do today to see if we could work on something together in maths?
- What could I do this week to improve the maths teaching in my class?
- Does everyone have a copy of our current policy and a scheme of work?
- How can I start to identify what people think is wanted and needed in our maths?
- Am I really applying what I know about managing people and the process of change?

APPENDICES

JOB DESCRIPTION FOR A MATHS CO-ORDINATOR

Job descriptions vary from school to school, but something along these lines might give a starting point for discussion.

1   Begin to implement a scheme of work for maths in the school by:

  • liaising with advisory service;
  • liaising with staff;
  • attending courses.

These activities will be undertaken as necessary.

2   Reorganise the resources by:
  • liaising with staff and making an inventory of teacher's handbooks, worksheets, etc.;
  • collecting and arranging these centrally according to the needs of the staff.

ACTION PLAN I

For action.

Name ....   Date ....

As a result of today I have identified these things that I need to do.

Of these things I will do .... first.

ACTION PLAN 2

My list for action

As a result of today, I have identified the following things that I need to do.

ACTION PLAN 3

Name ....

Date ....

Following the discussion about .... my plans are:

Short term

Medium term

Long term

ACTION PLAN 4

Name .... Date ....

Following the discussion on ...., my action plan is:

Aims

ACTION PLAN 5

Whole school action plan

Target

Those involved in working to achieve this target.

Tasks necessary to achieve this target.

-
-
-
-
-
-

Support required (e.g. INSET days, release time, professional development days, resources).

....

Development due to start on ....

Trying to complete by ....

Reviews will occur ....

Report to governors by ....

Thanks to Barbara MacGilchrist and Shirley Clarke for help with these action sheets for INSET. If you find this kind of action sheet has helped with your INSET, you can find more information in *The management of teaching and learning* by Barbara MacGilchrist and Shirley Clarke (see resources list).

REFERENCES AND RESOURCES

REFERENCES

Atkinson, S. (1992) *Mathematics with reason*, London, Hodder and Stoughton

Bird, M. (1991) *Mathematics for young children*, London, Routledge

Bounce to it and other books from Manchester Metropolitan University (see addresses below)

Briggs, R. (1973) *Jim and the beanstalk*, London, Puffin

Bruce, T. (1987) *Early childhood education*, London, Hodder and Stoughton

Campbell, R. J. (1985) *Developing the primary school curriculum*, London, Holt, Rinehart and Winston.

Chester, M. (no date) *Mathshare* (see addresses)

Cockcroft Report (1982) *Mathematics counts: report of the committee of inquiry into the teaching of mathematics*, HMSO, DES

Curriculum in Action: an approach to evaluation (1980) Milton Keynes, Open University Press

Dearing, R. (1994) *The National Curriculum and its assessment: final report*, London, Schools Curriculum and Assessment Authority

Denvir, B. and Brown, M. (1986) *Understanding of mathematical concepts in low-attaining 7–9 year olds*, Educational Studies in mathematics education, vol. 17 pp. 143–164

DES (1985) *Mathematics from 5–16*, Curriculum Matters 3, an HMI survey, HMSO

Donaldson, M. (1978) *Children's minds*, London, Fontana Collins

Floyd, A. (1981) *Developing Mathematical Thinking*, London, Addison-Wesley, publishers for the Open University

Fullan, M. (1982) *The meaning of educational change*, Toronto, OISE/Teachers College Press

Hughes, M. (1986) *Children and number*, Oxford, Blackwell

Hume, B. and Barrs, K. (1988) *Maths on Display*, Twickenham, Belair Publications

Liebeck P. (1984) *How children learn mathematics: a guide for parents and teachers*, London, Penguin

Lockhart, J. and Young, T. (1941) *Sure Foundation Arithmetic* book 3, University of Glasgow Press

McIntosh, A. (1981) 'When will they ever learn?' in Floyd, A. *Developing mathematical thinking*, London, Open University Press

OFSTED (1993) *The teaching and learning of number in primary schools*, London, HMSO

OFSTED (1994) *Science and mathemtics in schools*, London, HMSO

Open University (1982) *Developing mathematical thinking*, Floyd, A. et al., Milton Keynes, Open University Press

Shuard, H. (1986) *Primary mathematics today and tomorrow*, SCDC, Longman

Shuard, H. et al. (1990) *Children, mathematics and learning*, London, Simon and Schuster. (This large folder, produced by the many teachers that worked on the PrIME project, is a valuable resource for INSET and for working with parents.)

Shuard, H., Walsh, A., Goodwin, J. and Worcester, V. (1991), *PrIME Calculators, Children and Mathematics*, London, Simon and Schuster

CLASSROOM RESOURCES

BEAM   A wide range of excellent resources, including early years packs. These are mainly teacher's books, but include photocopiable sheets. These activities are written and trialled by teachers. Send for the brochure (see addresses below).

Bolt, B. (1982) *Mathematical Activities: A resource book for teachers*, Cambridge University Press

Bolt, B. (1985) *More Mathematical Activities*, Cambridge University Press

Burns, M. (1975) *The I Hate Mathematics! Book*, Cambridge University Press

Burton, L. (1984) *Thinking Things Through: Problem Solving in Mathematics*, Oxford, Blackwell

Deboys, M. and Pitt, E. (1980) *Lines of development in primary maths*, Belfast, Blackstaff Press

Exploring mathematics with young children and many other resources, ATM (see addresses below)

Gerrard, W. (1986) *I don't know – let's find out,* Leicester, Mathematical Association

Hume, B. and Barrs, K. (1988) *Maths on display,* Twickenham, Belair

Manchester Metropolitan books, *Bounce on it, Leap to it, Puzzle cards* and many others including strategy games (see address below)

Snape, C. and Scott, H. (1991) *How puzzling,* Cambridge University Press

Straker, A. (1993) *Talking points in mathematics,* Cambridge University Press

Tarquin products (see address below)
The catalogue is full of interesting maths resources and strategy games.

Williams, E. and Shuard, H. (1982) *Primary mathematics today,* London, Longman
Contains everything you need to know about the 'content' of maths.

Woodman, Ann et al. (1988) *Mathematics through art and design,* London, Unwin Hyman

Mathematical rhymes

Seven dizzy dragons (in press) Atkinson, S. et al., Cambridge University Press

This little puffin (new edition 1991) Matterson, E., London, Puffin

Count me in (1984) London, A. and C. Black.

Teaching maths from stories

Teach maths from stories (in press) Atkinson, S. et al., Cambridge University Press

Scholastic Child Education Maths Club, (see addresses)

Assessment

Clarke, S. and Atkinson, S. (1996) *Tracking significant achievement in primary maths,* London, Hodder and Stoughton

Drummond, M.J. (1993) *Assessing children's learning*, London, David Fulton

Starting from talking and thinking

Athey, C. (1990) *Extending thought in young children*, London, Paul Chapman

Durkin, K. and Shire, B. (1991) *Language in Mathematical Education*, Open University Press

Mathematics Association. (1987) *Maths talk*, Stanley Thornes

Ball, G. (1990) *Talking and learning*, Oxford, Blackwell

Brissenden, T. (1988) *Talking about mathematics*, Oxford, Blackwell

SAPERE (The Society for the Advancement of Philosophical Enquiry and Reflection in Education) c/o Roger Sutcliffe, Stammerham North, Christ's Hospital, Horsham, RH13 7NG.

If you are interested in getting your children talking and reasoning, SAPERE has a number of resources to get you started. These are not specifically mathematical but can be used as cross-curricular resources.

How do children learn maths?

Dickson, L., Brown, M. and Gibson, O. (1984) *Children learning mathematics: a teacher's guide to recent research*, Holt Rinehart and Winston for Schools Council

Pimm, D. (ed.) (1988) *Mathematics, teachers and children*, London, Hodder and Stoughton/Open University

Pimm, D. and Love, E. (eds) (1991) *Teaching and learning school mathematics*, London, Hodder and Stoughton

Skemp, R. (1986) 2nd edition *The psychology of learning mathematics*, London, Penguin

Skemp, R. (1989) *Mathematics in the Primary School*, London, Routledge

Tizard, B. and Hughes, M. (1984) *Young children learning*, London, Collins/Fontana

Vygotsky, L. (1978) *Mind in society*, Cambridge MA, Harvard University Press

The maths co-ordinator role

SUPPORT FROM GROUPS AND ORGANISATIONS

Join a BEAM group (write to BEAM for details).
Join ATM or MA (see addresses below).
Do a maths course. Contact your local university or write to the
Open University or BEAM.

BOOKS THAT CAN HELP

Campbell, R. J. (1985) (as above)

Easen, P. (1985) *Making school centred INSET work*, Routledge and
Open University
(Not specifically about maths but useful in any curricular area.)

Compton, G. and Davies, B. (no date) *Co-ordinating maths in
primary and middle schools*, ATM (see address below)

MacGilchrist, B. and Clarke, S. *School managed primary INSET
materials, Managing teaching and learning in school.* Available from
the Institute of Education, London (see addresses).

Open University (1990) *Working together: school based professional
development in mathematics* and *Working with colleagues*, in the
'Supporting Primary Mathematics' Pack
(Write to the OU centre for maths education, or ring 01908
653550.)

Pinner, M. and Shuard, H. (1984) *In-service education in primary
mathematics*, Open University Press

Stow, M. and Foxman, D. (1988) *Mathematics co-ordination: a study
of practice in primary and middle schools*, Windsor NFER/Nelson

Home/school resources

Scholastic now publish the IMPACT activities. These books are
photocopiable activities that can be taken home. The books are
divided into Key Stages 1 and 2 and cover measuring, shape and
space, number, etc.

Cambridge University Press publish Homelinks book as a part of
New Cambridge Mathematics.

There are a number of books available in high street shops that
parents can buy, such as the *Maths Challenge* books by David
Kirkby, published by Walker books.

A book for parents to read to help them to do some maths at home with their children is *Help your child with maths* by Sue Atkinson published by Hodder and Stoughton.

Parents might also like to send for the Tarquin catalogue.

USEFUL ADDRESSES

ATM (Association of Teachers of Mathematics), 7 Shartesbury Street, Derby, DE3 8YB

BEAM (Be A Mathematician), Barnsbury Complex, Offord Road, London, N1 1QH

Centre for maths education, Open University, Walton Hall, Milton Keynes, MK7 6AA or ring 01908 653550

Institute of Education, London University, 20 Bedford Way, London, WC1H 0AL

MA (Mathematical Association), 259 London Road, Leicester, LE2 8BE

Scholastic Publications Ltd., Villiers House, Clarendon Avenue, Leamington Spa, Warwickshire, CV32 5PR

Tarquin Publications, Stradbroke, Diss, Norfolk, IP21 5JP

A scheme of work called '*Mathshare*' is available, in either Key Stage 1 or a combined Key Stage 1 and 2, from Muriel Chester, 11 Fairacres, Bardolph Avenue, Croydon, CR0 9JY. It is available as a ring back A4 file with plenty of space for your own notes and references so that schools can use it with whatever resources they have available.